tredition

tredition was established in 2006 by Sandra Latusseck and Soenke Schulz. Based in Hamburg, Germany, tredition offers publishing solutions to authors and publishing houses, combined with world-wide distribution of printed and digital book content. tredition is uniquely positioned to enable authors and publishing houses to create books on their own terms and without conventional manu-facturing risks.

For more information please visit: www.tredition.com

TREDITION CLASSICS

This book is part of the TREDITION CLASSICS series. The creators of this series are united by passion for literature and driven by the intention of making all public domain books available in printed format again - worldwide. Most TREDITION CLASSICS titles have been out of print and off the bookstore shelves for decades. At tredi-tion we believe that a great book never goes out of style and that its value is eternal. Several mostly non-profit literature projects pro-vide content to tredition. To support their good work, tredition donates a portion of the proceeds from each sold copy. As a reader of a TREDITION CLASSICS book, you support our mission to save many of the amazing works of world literature from oblivion. See all available books at www.tredition.com.

 Project Gutenberg

The content for this book has been graciously provided by Project Gutenberg. Project Gutenberg is a non-profit organization founded by Michael Hart in 1971 at the University of Illinois. The mission of Project Gutenberg is simple: To encourage the creation and distribu-tion of eBooks. Project Gutenberg is the first and largest collection of public domain eBooks.

Popular Lectures on Zoonomia Or The Laws of Animal Life, in Health and Disease

Thomas Garnett

Imprint

This book is part of TREDITION CLASSICS

Author: Thomas Garnett
Cover design: Buchgut, Berlin – Germany

Publisher: tredition GmbH, Hamburg - Germany
ISBN: 978-3-8472-2318-4

www.tredition.com
www.tredition.de

Copyright:
The content of this book is sourced from the public domain.

The intention of the TREDITION CLASSICS series is to make world literature in the public domain available in printed format. Literary enthusiasts and organizations, such as Project Gutenberg, worldwide have scanned and digitally edited the original texts. tredition has subsequently formatted and redesigned the content into a modern reading layout. Therefore, we cannot guarantee the exact reproduction of the original format of a particular historic edition. Please also note that no modifications have been made to the spelling, therefore it may differ from the orthography used today.

POPULAR LECTURES ON ZOONOMIA, OR THE LAWS OF ANIMAL LIFE, IN HEALTH AND DISEASE.

BY THOMAS GARNETT, M.D.
MEMBER OF THE ROYAL COLLEGE OF PHYSICIANS, LONDON; OF THE ROYAL
IRISH ACADEMY; OF THE ROYAL MEDICAL SOCIETY OF EDINBURGH; HONORARY
MEMBER OF THE BOARD OF AGRICULTURE; FELLOW OF THE LINNEAN SOCIETY;
MEMBER OF THE MEDICAL SOCIETY, LONDON; AND OF THE LITERARY AND
PHILOSOPHICAL SOCIETY OF MANCHESTER: &c. &c.
FORMERLY PROFESSOR OF NATURAL PHILOSOPHY AND CHEMISTRY IN THE
ROYAL INSTITUTION OF GREAT BRITAIN.

LONDON:
FROM THE PRESS OF THE ROYAL INSTITUTICN OF GREAT BRITAIN:
W. SAVAGE, PRINTER.
PUBLISHED FOR THE BENEFIT OF THE AUTHOR'S CHILDREN BY HIS EXECUTORS.
TO BE HAD OF MR. NICHOLSON, SOHO SQUARE, MR. PRICE, WESTMINSTER LIBRARY, JERMYN STREET,
AND OF ALL THE BOOKSELLERS.
1804.

[FRONTISPIECE PORTRAIT]

THOMAS GARNETT. M.D.

L. R. Smith, del.
Lenney, sculpt.

Published Jan. 1, 1805, by the Executors, for the benefit of his orphan children.

ENTERED AT STATIONERS HALL.

TO THE RIGHT HONOURABLE, AND HONOURABLE, THE MANAGERS OF THE ROYAL INSTITUTION OF GREAT BRITAIN, THESE LECTURES, COMPOSED BY A MAN, WHO, IN HIS LIFE TIME, WAS HONOURED BY THEIR SELECTION, AS THEIR FIRST LECTURER; AND WHOSE INFANT FAMILY HAVE SINCE EXPERIENCED THEIR BENEVOLENCE AND PROTECTION, ARE, WITH PERMISSION, DEDICATED, BY THE TRUSTEES OF THE SUBSCRIPTION, IN FAVOUR OF THOSE ORPHANS

CONTENTS.

THE AUTHOR'S LIFE.

His early amusements. His apprenticeship to Mr. Dawson. His studies
at Edinburgh. In London. His establishment at Bradford. At
Knaresborough. At Harrowgate. His marriage. His lectures at
Liverpool. At Manchester. At Warrington. At Lancaster. At Glasgow.
His tour in the Highlands. The death of his wife. His engagement in
the Royal Institution. His resignation. His establishment in
Marlborough Street. His appointment as physician to the Mary-le-bonne
Dispensary. His death.

LECTURE I, INTRODUCTION.

Difficulties and advantages of a popular course of lectures. General view of the human frame. Bones. Muscles. Joints. Powers of the muscles. Brain and Nerves. Senses. Hypothesis of sensation. Galvanism. Distribution of the subjects of the course.

LECTURE II, ON RESPIRATION.

Air. Trachea. Thorax. Animal heat. Its uniformity. Chemical properties of the air. Combustion. Effects of cold.

LECTURE III, ON THE CIRCULATION OF THE BLOOD.

Respiration partially voluntary. Heart. Circulation. Pulsation. Hepatic vessels. Action of the arteries. Causes propelling the blood. Varieties of the pulse. Changes of the blood. Harvey's merits.

LECTURE IV, ON DIGESTION AND NUTRITION.

Necessity of food. Structure of the viscera. Bile. Food of man. Gastric juice. Absorption. Assimilation. Lymphatics. Diseases affecting digestion. Advantages of temperance and exercise.

LECTURE V, OF THE SENSES IN GENERAL.

Sensation. Attention. Internal senses. Habit. Touch. Skin. Pain.

LECTURE VI, ON TASTE AND SMELL.

Tongue. Kinds of taste. Diseases of taste. Smell. Mucous membrane. Odours. Smell in animals. Diseases of smell.

LECTURE VII, ON SOUND AND HEARING.

Production of sound. Medium. Ear. Hearing. Pendulums. Chords. Wind instruments. Tones. Velocity of sound. Music. Echo. Deafness.

LECTURE VIII, ON VISION.

The eye. Figure. Light. Vision. Accommodation to different distances.
Seat of vision. Erect vision. Single vision. Squinting.

LECTURE IX, ON THE LAWS OF ANIMAL LIFE.

Action of external objects. Excitability. Its laws. Action of light. Of Heat. Of food. Sound. Odours.

LECTURE X, ON THE LAWS OF ANIMAL LIFE.

General laws. Sleep. Degrees of excitability. Health. Comparison with a furnace. Oxidation. Electricity. Hydrogen. Theory of muscular contraction.

LECTURE XI, OF THE NATURE AND CAUSES OF DISEASES.

Brown's theory. Sthenic and asthenic diseases. Debility. Sthenic depression of spirits. Scale of excitability. Fallacy of symptoms Effects of cold. Alcohol. Sthenic diseases.

LECTURE XII, ON INFLAMMATION AND ASTHENIC DISEASES.

Nature of inflammation. Distention of the arteries. Cure of ophthalmias. Asthenic diseases. Cold. Intemperance. Mental exertions. Classes of diseases. Cure. Oxidation.

LECTURE XIII, ON THE GOUT.

Effects of the gout. Gout not hereditary. Symptoms. Causes. Affections of the stomach. Cure. Use of electricity. Diet.

LECTURE XIV, ON NERVOUS COMPLAINTS.

Predisposition. Classes. Sthenic kinds. Case of the author. Bad effects of wine. Asthenic kinds. Passions. Direct debility. Treatment. Torpor. Remedies. Exercise and temperance. Conclusion.

AN ACCOUNT OF THE LIFE OF THE AUTHOR

DR. GARNETT was born at Casterton, near Kirkby Lonsdale, Westmoreland, on the 21st of April, 1766. During the first fifteen years of his life, he remained with his parents, and was instructed by them in the precepts of the established church of England, from which he drew that scheme of virtue, by which every action of his future life was to be governed. The only school education he received during these early years, was at Barbon, a small village near his native place, to which his father had removed the year after he was born. The school was of so little consequence, that its master changed not less than three times during the space of seven or eight years, and the whole instruction he received, was comprehended in the rudiments of the English grammar, a small portion of Latin, and a little French, together with the general principles of arithmetic. His bodily constitution was from the beginning weak and susceptible; he was unequal to joining in the boisterous amusements of his companions, while from the liveliness of his disposition he could not remain a moment idle. To these circumstances we are, perhaps, to attribute the uncommon progress he made in every branch of knowledge to which he afterwards applied himself.

Whilst a schoolboy, the susceptibility of his mind, and a diffidence of character connected with it, caused him to associate very little with his schoolfellows: he dreaded the displeasure of his preceptor, as the greatest misfortune which could befal him The moment he arrived at home, he set about preparing his lesson for the next day; and as soon as this was accomplished, he amused himself by contriving small pieces of mechanism, which he exhibited with conscious satisfaction to his friends. His temper was warm and enthusiastic; whatever came within the narrow circle of his early knowledge he would attempt to imitate. He saw no difficulties before hand, nor was he discouraged when he met with them. At the early age of eleven years, he had somewhere seen a dial and a quadrant, and was able to imitate these instruments, nay, with the assistance of the latter, and the small knowledge of arithmetic and trigonometry, which he had then obtained, he formally marched out with his younger brother, and rudely attempted to measure the height of a mountain behind his father's house. When he was nearly fifteen years of age, he was, at his earnest desire, put apprentice to the celebrated mathematician, Mr. Dawson, of Sedbergh, who was at that time a surgeon and apothecary. This situation was peculiarly advantageous to him, on account of the great mathematical knowledge of his master, by whom he was instructed in the different branches of this science; and, notwithstanding his constant employment in necessary business, his ardent pursuit of professional information, and his extreme youth, in the course of four years, he became well acquainted with mechanics, hydrostatics, optics, and astronomy. He afterwards applied himself with energy to the study of chemistry, and other subjects, with which it was thought expedient that he should be acquainted, previously to attending the medical lectures in the University of Edinburgh. Strongly impressed with a sense of the value of time, he was indefatigable in the pursuit of knowledge: by a concurrence of fortunate circumstances, his talents had become so flexible, that he succeeded almost equally well in every subject to which he applied himself; but of chemistry he was particularly fond, and from this time it became his favourite study.

During the four years of his apprenticeship, his conduct was in every respect highly commendable; he was assiduous, he was virtuous. His pursuit after general knowledge was restrained to one object only at a time; he had advanced far in the abstruse sciences; his inclination for study was increased: when in the year 1785, he went to Edinburgh with a degree of scientific knowledge, seldom attained by young men beginning the study of medicine. He became a member of the Medical and Physical Societies, where he soon made himself conspicuous, and of the latter of which, he was afterwards president.

Well acquainted with the first principles of natural philosophy, he had considerable advantages over his contemporaries; and his superiority was soon acknowledged. He was not, however, on this account inclined to remit his industry; he attended the lectures of the ablest professors of the day, and more particularly those of Dr. Black, with the most scrupulous punctuality, and endeavoured to elucidate his subject by every collateral information he could obtain. He avoided almost all society; and it is said, he never allowed himself, at this time, more than four hours sleep out of the twenty four. The famous Dr. Brown was then delivering lectures on his new theory of medicine. Dr. Garnett, fired with the enthusiasm of this noted teacher, and struck with the conformity of his theory to the general laws of nature, became one of the most zealous advocates of his doctrine; and from this period, he took, during the remainder of his life, every opportunity of supporting it.

During two summers he returned to Mr. Dawson at Sedbergh, passing the intervening winters in Edinburgh: about this time he wrote the essay, which, in the year 1797, he published under the title of a Lecture on Health, which very neatly and perspicuously explains the fundamental parts of the Brunonian theory of medicine: in September 1788, he published his inaugural dissertation de Visu, and obtained the degree of M.D. Very soon afterwards he went to London, to pursue his professional studies, which he continued to do with the greatest perseverance: he attended with unceasing diligence the lectures of the most eminent lecturers, and he sought practical knowledge in the chief hospitals of the metropolis with the most ardent zeal; so that whilst he gained information to himself, he set an impressive example to his contemporary medical students, who in the delusive pursuits of a great city, are too apt to neglect the objects their parents had in view in sending them to the capital. Having finished his studies in London, Dr. Garnett, in 1789, returned to his parents. At the time he left London, he had lost none of his ardour; still he continued indefatigable and observant. He had been flattered and respected by his fellow students, and praised by his seniors; and his previous success animated him with the strongest expectation of future advancement. At this time, it is supposed, he wrote the justly admired Treatise on Optics, which is in the Encyclopaedia Britannica. Soon after his establishment as a physician, at Bradford, in Yorkshire, which took place in the year 1790, he began to give private lectures on philosophy and chemistry. He wrote his treatise on the Horley Green Spa; and in a short time, gained a deserved character of ingenuity and skill as a chemist, a physician, and a benevolent member of society. Bradford did not afford scope for his practice as a physician, equal to the sanguine expectations he had formed; and he was induced to change his situation.

In the year 1791, therefore, he removed to Knaresborough, intending to reside at that place during the winter, and at Harrowgate during the summer. This plan he put in execution till the year 1794; his reputation rapidly increased, and his future prospects appeared cheering and bright. He continued to apply himself very closely to chemistry, which was now decidedly his most pleasant and interesting study. He endeavoured to apply his various knowledge to practical purposes, and in many instances was peculiarly successful. No sooner had he arrived at Knaresborough, than anxious to investigate every thing in the neighbourhood, which could at all affect the health of the inhabitants, he began to analyse the Crescent Water at Harrowgate; which he did, with all the accuracy a subject so difficult could admit of; and in 1791, he published his treatise upon it. The same spirit led him, in 1792, to analyse the other mineral waters at the same place of fashionable and general resort, the detail of which he published in the same year. These publications became generally read, and gained him a very extensive reputation. The late Dr. Withering, whose knowledge on these subjects could not be disputed, before he had seen his general analysis of the Harrowgate Waters, said, that "excepting only the few examples given us by Bergman, the analysis of the Crescent Waters was one of the neatest and most satisfactory accounts he had ever read of any mineral water." But his exertions were not confined to professional and scientific pursuits; laudably desirous of advancing knowledge amongst every branch of the community, he formed the plan of a subscription library, which has, since 1791, been of great convenience and utility to the inhabitants of Knaresborough. Far from joining in the opinion which has so much prevailed in modern times, that it was sufficient to aim at general utility, he lost no opportunity of doing good to every member of society. He greatly promoted and encouraged the making of the pleasure grounds and building on the rock, called Fort Montague; and he instructed and assisted the poor man, who is called the Governor, to institute a bank, and to print and issue small bills of the value of a few halfpence, in imitation of the notes of the country bankers, but drawn and signed with a reference of humour to the fort, the flag, the hill, and the cannon. These notes, the nobility and gentry, who during the Harrowgate season crowd to visit this remarkable place, take in exchange for their silver, and by these means the governor, who is a man of gentle and inoffensive manners, has been enabled, with the assistance of his loom, to support himself and a numerous family, and to ameliorate their condition, by giving education to his children.

No station in life escaped his benevolent attentions. In order to benefit John Metcalf, who is perhaps more generally known by the name of Blind Jack of Knaresborough, he assisted him to publish an account of the very singular and remarkable occurrences of his life, during a long series of years, under the heavy affliction of total blindness; by the sale of which, this venerable old man derived a considerable contribution towards his subsistence

Whilst at Harrowgate, Dr. Garnett obtained the patronage and protection of the Earl of Rosslyn, then Lord Loughborough, who in the year 1794 built a house for him, which for the future Dr. Garnett meant should be his only residence; it was not long however before he discovered that his situation at Harrowgate was but ill calculated to forward his liberal and extended views. At this place he had small opportunities of attaching himself to his favourite sciences; in the winter months he was without literary society, and it was not for his ardent spirit to remain inactive. About this time also, he formed the idea of going to America, where he thought he might live both honourably and profitably as a teacher of chemistry and natural philosophy. All these circumstances were floating in his mind, when in the year 1794, about the end of July, at the instance of a medical friend, who resided in London, he received as boarders into his house, which was kept by his sister, Miss Catharine Grace Cleveland, daughter of the late Mr. Cleveland, of Salisbury Square, Fleet Street, who was recommended to the use of the Harrowgate waters, together with her friend Miss Worboys. To all who were acquainted with the prepossessing exterior of Dr. Garnett, the liveliness of his conversation, the urbanity of his manners, and his general desire of communicating knowledge to whomever he saw desirous of gaining information, it will be no surprise, that a mutual attachment grew up between him and his inmate, Miss Cleveland, a young lady possessing, in all respects, a mind similar to his own, and who must have felt a natural gratification in the zeal with which the company of the person, on whom she had placed her affections, was sought by all ranks resorting to this fashionable watering place, where every one thought himself most fortunate who sat nearest to him at the table, and where he enlivened the circle around him with his conversation, which was not only instructive, but playfully gay, and entertaining, ever striving to amuse, and always successful in his attempts. The Doctor now began to project plans of happiness, which he had only before held in idea. Previous to his visitors leaving Harrowgate, which was towards the latter end of December, he communicated to Miss Cleveland his intention of going to America. At first she hesitated about accompanying him; but finding his resolution fixed, she at length consented. From this time, till the beginning of March 1795, he continued deliberating upon and maturing his plan. He now departed from Harrowgate, and followed the object of his affection to her mother's residence at Hare Hatch, Berks. He was married to her on the 16th of March, and a fortnight afterwards returned to Harrowgate, to dispose of the lease of his house, and his furniture. Having again joined his wife, he then went to London, where he purchased apparatus for his lectures, and after visiting his parents, he proceeded to Liverpool, in order to obtain a passage to America.

Whilst he was thus waiting for the opportunity of a vessel to transport him across the Atlantic, he was solicited by the medical gentlemen at Liverpool, to unpack his apparatus, and give a public course of lectures on chemistry and experimental philosophy. At all times desirous of diffusing the knowledge he had acquired, and eager to fulfil the wishes of his friends, he complied with their request, and entered upon a plan, which in the end completely overturned the scheme he had for several months been contemplating with such ardent hopes of happiness and prosperity. No sooner had he been prevailed upon, than he set about getting every thing ready for his lectures, and after a single week's preparation; he commenced his course. The deep interest he took in his subject, the anxiety he showed to make himself understood, and the enthusiastic hope he constantly expressed of the advancement of science, had a remarkable effect upon his audience; and his lectures were received with the most flattering marks of attention, and excited the most general applause and satisfaction. In a short time, he received a pressing invitation from the most eminent characters at Manchester, to repeat his course in that town. This invitation he accepted, and, encouraged by the success he had just experienced, he postponed the idea of leaving his country. He arrived at Manchester about the middle of January 1796, and began his lectures on the 22nd of that month. Before his arrival, not less than sixty subscribers had put down their names, the more strongly to induce him to comply with their wishes, and many more had promised to do it, as soon as his proposals were published. Notwithstanding he was thus led to expect a large audience, and had procured apartments, which he imagined would be sufficiently spacious for their reception, he was obliged, for want of room, to change them not less than three times during one course. With such success did the career of his philosophical teaching begin, and with such extreme attention and respect was he every where received, that he used afterwards to mention this period, as not only the most profitable, but the most happy of his life. On the 24th of February, his wife was brought to bed of a daughter, the eldest of the two orphans who have now to lament the death of so valuable a parent, to deplore the loss of that independence which his exertions were certain to have raised them, and to rely on a generous public for protection, in testimony of the virtues and merit of their father.

After this time Dr. Garnett repeated nearly the same course of lectures at Warrington and at Lancaster; to both which places he was followed by the same success.

Whilst he was in this manner exerting himself for the general diffusion of knowledge, his fame spread with the delight and instruction he had every where communicated to his audience. The inhabitants of Birmingham wished to have the advantage of his lectures; and he also received a most pressing invitation from Dublin, where a very large subscription had already been formed. It was his intention to have accepted of the latter invitation, but previous to his departure for Ireland (from whence he had even yet some thoughts of emigrating to America) he was informed of the vacancy of the professorship in Anderson's Institution, at Glasgow, by his friend the late Dr. Easton of Manchester, who strongly urged him to become a candidate. As this situation must inevitably destroy all his future prospects, he for a long time hesitated; but Dr. Easton having informed the Managers of the Institution, that there was a possibility of their obtaining a professor, so eminently qualified as Dr. Garnett, they, after making further inquiry concerning him, offered it to him in so handsome a manner, that, although the situation was by no means likely to be productive of so much emolument as the plan of life he had lately been pursuing, he yielded to their proposal, strengthened as it was by the earnest solicitation of Mrs. Garnett, who felt considerable apprehension at the thoughts of going to America, and consented to accept of the professorship.

He began his lectures at Glasgow in November 1796, and a short account of them may be found in his Tour to the Highlands, vol. ii. p. 196. The peculiar clearness with which he was wont to explain the most difficult parts of science, together with the simplicity of the terms he employed, rendered his lectures particularly acceptable to those who had not been initiated in the technical terms, generally used on such occasions. Every thing he delivered might easily be understood by those who had not previously attended to the subject; and of consequence, all who had been disgusted, or frightened by the difficulties they had before met with, or imagined, were eager to receive his instructions; and the audience he obtained, was much more numerous, than either the trustees, or himself, had deemed probable.

When the session was completed, he repaired to Liverpool for the purpose of fulfilling a promise he had formerly given to his friends, to repeat his course of lectures in that town. Mrs. Garnett, in the mean time, remained at Kirkby Lonsdale, where he joined her as soon as his lectures were finished. He spent the latter part of the summer chiefly in botanical pursuits, and returned to Glasgow in the autumn, when he made known his intention of practising as a physician. Fortune continued to favour him, his reputation increased, and he rapidly advanced towards the first professional situation in Glasgow.

In July 1798, he began his Tour to the Highlands, an account of which he published in 1800, and having returned to his duties in the Institution, the success of his lectures suffered no interruption, but whilst he was reaping the benefit due to his industry and his talents, his happiness received a blow, which was irrecoverable, by the loss of his wife, who died in child birth, December the 25th 1798: the infant was preserved. The sentiments of Dr. Garnett on this occasion will be best expressed in his own words, in a letter to Mr. Ort, of Bury in Lancashire.

"Glasgow, January 1st. 1799.

"Oh my dear cousin, little did I expect that I should begin the new year with telling you that I am now deprived of all earthly comforts; yes, the dear companion of my studies, the friend of my heart, the partner of my bosom, is now a piece of cold clay. The senseless earth is closed on that form which was so lately animated by every virtue; and whose only wish was to make me happy.

"Is there any thing, which can now afford me any consolation? Yes, she is not lost, but gone before: but still it is hard to have all our schemes of happiness wrecked: when our bark was within sight of port, when we were promising ourselves more than common felicity, it struck upon a rock: my only treasure went to the bottom, and I am cast ashore, friendless, and deprived of every comfort. My poor, dear love had been as well as usual during the two or three last months, and even on the dreadful evening (christmas eve) she spoke with pleasure of the approaching event. My spirits were elevated to so uncommon a pitch, by the birth of a lovely daughter, that they were by no means prepared for the succeeding scene; and they have been so overwhelmed, that I sometimes hope it may be a dream, out of which I wish to awake. The little infant is well, and I have called it Catharine, a name which must ever be dear to me, and which I wish to be able to apply to some object whom I love; for though it caused the death of my hopes, it is dear to me, as being the last precious relic of her, whom every body who knew her esteemed, and I loved. I must now bid adieu to every comfort, and live only for the sweet babes. Oh! 'tis hard, very hard. "THOMAS GARNETT." "To Mr. Ort, Bury, Lancashire.

The affliction Dr. Garnett experienced on the death of his wife, was never recovered. On all occasions of anxiety which were multiplied upon him, by reason of his exquisite sensibility, he longed for the consolation her society used to afford him; and although his susceptibility to the action of external causes, would not allow him to remain in continued and unalterable gloom and melancholy, yet in solitude, and on the slightest accident, his distress returned, and he despaired of the possibility of ever retrieving his lost happiness. Had it not been for his philosophical pursuits, and the duties of his extensive practice, which kept him almost constantly engaged, it may be doubted, whether he could at this time have sustained the load of sorrow with which he was oppressed.

The circumstances which remain to be mentioned are few. From the death of his wife, Dr. Garnett may be considered as unfortunate; for although a fair prospect opened before him, a series of occurrences took place, which neither his state of mind, nor his constitutional firmness enabled him to support.

At the time of the formation of the Royal Institution of Great Britain, in London, Count Rumford wrote to Dr. Garnett, to whom he was then an entire stranger, inquiring into the nature and economy of Anderson's Institution, Glasgow; the plan of the lectures given, &c. &c.; and after hinting at the opportunities of acquiring reputation in London, he finally proposed that Dr. Garnett should become lecturer of the new Institution. With this proposal, arduous as was the task, to deliver a course of lectures on almost every branch of human attainment, Dr. Garnett complied, relying on his acquirements, and the tried excellence of his nature; and conscious that no difficulty could resist the indefatigable exertions which on other occasions he had so successfully applied. Flattered by the honour and respect he conceived to be paid to his abilities and qualifications; pleased with the prospect of more rapidly accumulating an independence for himself and his children; and animated with the hope of meeting with more frequent opportunities of gratifying his thirst after knowledge, his spirits were again roused, and he looked forward to new objects of interest in the advancement of his favourite pursuits. In the enthusiasm of the moment, he was known to say, that he considered his connexion with the Royal Institution, from which the country had a right to expect so much, as one of the most fortunate occurrences of his life. On the 15th October 1799, he informed a special meeting of the Managers of Anderson's Institution, of his appointment to the Professorship of Philosophy, Chemistry, and Mechanics, in the Royal Institution of Great Britain, and on that account requested permission to resign his situation. The resignation of a man, whom all loved and revered, was reluctantly, though, as tending to his personal advancement, and the promotion of science, unanimously accepted by the meeting; he was congratulated on his new appointment, and thanked for the unremitting attention he had paid to the interests of Anderson's Institution, ever since he had been connected with it. As an instance of the high esteem in which he was held by the trustees, it may be observed, that his successor, Dr. Birkbeck, was elected by a very great majority of votes, principally on account of his recommendation. In November, he pursued his journey to London, leaving his children at Kirkby Lonsdale, under the care of Miss Worboys. This lady, whose friendship for Mrs. Garnett had induced her to become almost her constant companion, and had even determined her to go with her friend to America, if the Doctor had put his intentions in execution; soon after the death of Mrs. Garnett, had pledged herself, never to desert the children, so long as she could be of any use to them. How faithfully she observes this obligation, all who know her must acknowledge; nor can we, without increased anxiety, reflect upon the situation the poor orphans must have been in without her protection.

Dr. Garnett was received by the Managers of the Royal Institution with attention, civility, and respect. During the winter, the lecture room was crowded with persons of the first distinction and fashion, as well as by those who had individually contributed much to the promotion of science; and although the northern accent, which Dr. Garnett still retained in a slight degree, rendered his voice somewhat inharmonious to an audience in London, his modest and unaffected manner of delivering his opinions, his familiar, and at the same time elegant language, rendered him the object of almost universal kindness and approbation.

The exertions of the winter had in some measure injured his health, and a degree of uncertainty that he saw in his prospects, tended greatly to depress his spirits. He determined, however, to keep his situation at the Institution, in order that he might at a more convenient time be justified to himself in resigning it. In the summer of 1800, he visited his children in Westmoreland; but his anxiety of mind was not diminished, nor consequently his health improved, by this relaxation from active employment. He walked over the same ground, and viewed the same prospects that he had formerly enjoyed in the company of his wife. He had not resolution to check the impressions as they arose; and thus, instead of being solaced by the beauties which surrounded him, he gave the reins to his melancholy fancy, which, unchecked by any other remembrance, dwelt only on the affection and the virtues of her, whose loss he had ever to deplore; the want of whose society he imagined to be the chief source of his misery. Towards the end of autumn, he returned to the Institution, and in the winter, recommenced his duties as professor. The effect produced upon his lecturing by these and other irritating circumstances was remarkable. Debility of body, as well as uneasiness of mind, incapacitated him for that ardent and energetic pursuit of knowledge, by which he had been so eminently distinguished. His spirited, and at the same time modest method of delivery was changed into one languid and hesitating, that, during this period, occasioned an erroneous judgment to be formed of his abilities as a man of science, and a teacher, by such of his audience as were unacquainted with the cause, or the intrinsic value and merit of the man. At the close of the season, his determination of retiring from the Institution was fixed; and he presented to the Managers his resignation.

It was well known to Dr. Garnett's particular friends, that during the early part of this session, he determined to withdraw himself from the Institution; but the success and advancement of the establishment, which he sanguinely hoped would stand unrivalled in the universe, was so intimately connected with the affections of his mind, that he resolved to forego every personal consideration, rather than risk an inconvenience to the Institution, by ceasing to deliver his lectures in the middle of a course; liberally considering, that the Managers, after the business of the season was over, would have time and opportunity before the ensuing session, to fill the professor's chair with talents competent to the arduous undertaking; a circumstance the Managers afterwards so eminently profited by, with the highest credit to themselves, and advantage to the public, in the nomination of the gentlemen who now fill the situation held by Dr. Garnett, and who discharge its important duties with the most distinguished abilities.

The transactions of the last winter almost completely served to undermine the small strength of constitution he had left; he was constantly harrassed by complaints in the organs of digestion; head ache deprived him of the power of application; his countenance assumed a sallow complexion; the eye which had beamed with animation, retired within its socket, deprived of lustre; melancholy conceptions filled his imagination more habitually, and were excited by slighter causes; at times, they altogether deprived him of the power of exertion; and while he lamented their effect, the contemplation of themselves rendered him the more their prey. At this time, a gloomy day, or the smallest disappointment, gave him inconceivable distress; but he was not altogether incapable of temporary relief, and the few moments of pleasure he seemed to enjoy, would have given reason to believe, that he might once more have recovered, and have long continued to be the delight and instructor of his friends. A more close observation would at the same time have justified the supposition, that the strong and painful emotions of mind he had suffered, had already induced disorders of the bodily system, which were irrecoverable. Before Doctor Garnett had left his situation at Glasgow, he had determined to practice as a physician in London; but from this he was restrained, during the time he was at the Royal Institution. To his former intention he now determined to apply himself, and in addition to the attempt, by giving private lectures, to assure himself of that independency, of which his unfortunate destiny, though with every reasonable expectation before him, had hitherto deprived him.

With this intention, he purchased the lease of a house in Great Marlborough Street; and in the summer of 1801, built a lecture room. He brought his family to town, and once more looked forward with hope. The flattering success he soon met with, and a short residence at Harrowgate in the autumn, contributed to afford a temporary renovation of health and spirits; it was, however, but a short and delusive gleam of prosperity which now dawned upon him; for, confiding too much in his newly increased strength, he exerted himself to a much greater degree than prudence would have suggested. In the course of the following winter, he delivered not less than eight courses of lectures, two on chemistry, two on experimental philosophy, a private course on the same subject, one on mineralogy, and the course to which this sketch is prefixed, which he also delivered in an apartment at Tom's Coffee house, for the convenience of medical students, and others, in the city. Besides these, he commenced two courses on botany, one at Brompton, and the other at his own house; but a return of ill health prevented his concluding them. It was not to be expected, that a constitution so impaired and debilitated, could long support this continued labour of composition and recitation; accordingly he became affected with a consequent disorder, which rapidly exhausted his strength; and, being unable to employ the only probable means of recovering it, he became more incapable of exertion. His spirits however were roused, and he ceased not to use every means of increasing his practice. In the spring of 1802, the office of physician to the St. Mary le Bonne Dispensary happened to be vacant, and he became a candidate; he was more than commonly anxious to obtain this situation. It seemed to him, as if his future good or ill fortune depended altogether upon the event of his canvass, he spared no effort to ensure his success; and accordingly was appointed to the situation in May. His life now drew near a close. Little was he calculated to bear the accumulated labours, and extreme fatigue, to which he was daily exposed. Any benefit which might have resulted from constant and well regulated occupation was frustrated; for whilst he still suffered from the vividness of his conception, representing to him in mournful colours the occurrences of his past life, he became liable to other evils, not less injurious and destructive. The practice of medicine requires both vigorous health of body and firmness of mind. Dr. Garnett, now greatly weakened in body, and not exempt from anxiety of mind, became more and more susceptible to the action of morbific matter. It was not long before he received the contagion of typhous fever, whilst attending a patient, belonging to that very dispensary of which he had been so anxious to become physician. He laboured under the disorder for two or three weeks, and died the 28th of June, 1802; and was buried in the new burial ground of the parish of St. James, Westminster.

Thus was lost to society a man, the ornament of his country, and the general friend of humanity. In his personal attachments, he was warm and zealous. In his religion he was sincere, yet liberal to the professors of contrary doctrines. In his political principles, he saw no end, but the general good of mankind; and, conscious of the infirmity of human judgment, he never failed to make allowances for error. As a philosopher, and a man of science, he was candid, ingenuous, and open to conviction; he never dealt in mystery, or pretended to any secret in art; he was always ready in explanation, and desirous of assisting every person willing to acquire knowledge. Virtue was the basis of all his actions; science never possessed a fairer fabric, nor did society ever sustain a greater loss.

LECTURES ON ZOONOMIA.

LECTURE I. INTRODUCTION.

I AM well aware of the difficulties attending the proper composition of a popular course of lectures on the animal economy, which must be essentially different from those generally delivered in the schools of medicine; because it professes to explain the structure and functions of the living body, to those who are supposed to be unacquainted with the usual preliminary and collateral branches of knowledge. It must be obvious to every one, that it can be by no means an easy task to give in a few lectures, a perspicuous view of so extensive a subject; but I trust that the consideration of this difficulty will readily extend to me your indulgence.

That such a course, if properly conducted, must be interesting, needs scarcely to be observed; for the more we examine the structure and functions of the human body, the more we admire the excellence of the workmanship, and beauty of contrivance, which presents itself in every part, and which continually shows the hand of omniscience. The most ingenious of human inventions, when compared with the animal frame, indicate a poverty of contrivance which cannot fail to humble the pretensions of the sons of men. Surely then there are few who will not feel a desire to become acquainted with subjects so interesting.

But there is another point of view which will place the utility of such inquiries in a still stronger light. We shall afterwards see, that our life is continually supported by the action of a number of substances, by which the body is surrounded, and which are taken into the stomach for its nourishment. On the due action of these depends the pleasant performance of the different functions, or the state of health; without which, riches, honours, and every other gratification, become joyless and insipid.

By understanding the manner in which these powers act, or, in other words, by becoming acquainted with the principles of physiology, we shall be enabled to regulate them, so as, in a great measure, to guard against the numerous ills that flesh is heir to: for it is universally agreed, that by far the greatest part of the diseases to which mankind are subject, have been brought on by intemperance, imprudence, and the neglect of precautions, which often arises from carelessness, but much oftener from ignorance of those precautions.

Physiological ignorance is, undoubtedly, the most abundant source of our sufferings; every person accustomed to the sick must have heard them deplore their ignorance of the necessary consequences of those practises, by which their health has been destroyed: and when men shall be deeply convinced, that the eternal laws of nature have connected pain and decrepitude with one mode of life, and health and vigour with another, they will avoid the former and adhere to the latter.

It is strange, however, to observe that the generality of mankind do not seem to bestow a single thought on the preservation of their health, till it is too late to reap any benefit from their conviction: so that we may say of health, as we do of time, we take no notice of it but by its loss; and feel the value of it when we can no longer think of it but with retrospect and regret.

When we take a view of the human frame, and see how admirably each part is contrived for the performance of its different functions, and even for repairing its own injuries, we might at first sight imagine, that such a structure, unless destroyed by external force, should continue for ever in vigour, and in health: and it is by mournful experience alone that we are convinced of the contrary. The strongest constitution, which never experienced the qualms of sickness, or the torture of disease, and which seems to bid defiance to the enemies of health that surround it, is not proof against the attacks of age. Even in the midst of life we are in death; how many of us have contemplated with admiration the graceful motion of the female form; the eye sparkling with intelligence; the countenance enlivened by wit, or animated by feeling: a single instant is sufficient to dispel the charm: often without apparent cause, sensation and motion cease at once; the body loses its warmth, the eyes their lustre, and the lips and cheeks become livid. These, as Cuvier observes, are but preludes to changes still more hideous. The colour passes successively to a blue, a green, and a black; the flesh absorbs moisture, and while one part of it escapes in pestilential exhalations, the remaining part falls down into a putrid liquid mass. In a short time no part of the body remains, but a few earthy and saline principles; its other elements being dispersed through air, or carried off by water, to form new combinations, and afford food for other animals.

The human body has been defined to be a machine composed of bones and muscles, with their proper appendages, for the purpose of motion, at the instance of its intelligent principle. From this principle, nerves, or instruments of sensation, are likewise detached to the various parts of the body, for such information as may be necessary to determine it to those motions of the body, which may conduce to the happiness of the former, and the preservation of both.

It may perhaps be objected to this definition, that the body consists of other parts besides bones, muscles, and nerves; this is undoubtedly true; but, if we examine more minutely, we shall find that all the other parts, as well as functions of the body, seem only to be subservient to the purposes I have mentioned. For, in the first place, the muscles which are necessary to the motions of the body, are, from the nature of their constitution, subject to continual waste; to repair which waste, some of the other functions have been contrived.

Secondly, most of the other parts and functions of the body, are either necessary to the action of the muscles, or to the operation of the intelligent principle, or both.

Lastly, from the sensibility, and delicate structure, of the muscles and nerves, they require to be defended from external injuries: this is done by membranes, and other contrivances, fitted for the purpose.

To see this more clearly, we shall examine a little more particularly how each of the functions is subservient to the muscular and nervous systems. For this purpose it may be observed, 1st. that the stomach and digestive faculties serve to assimilate the food, or convert it into matter proper to repair the continual waste of solids and fluids. The circulation of the blood besides being absolutely necessary, as we shall afterwards see, to the action of the muscles, distributes the nourishment, thus assimilated and prepared by the stomach, to all parts of the body. The different glands separate liquors from the blood, for useful, but still for subservient purposes. Thus the salivary glands, stomach, pancreas, and liver, separate juices necessary to the proper digestion and assimilation of the food. The kidneys serve to strain off from the blood the useless and superfluous water, salts, &c. which if allowed to remain in the body would be very injurious to it.

We shall afterwards see, that the nerves are not only instruments of sensation, but the origin of motion; it being immediately by their means that the muscles are moved. A certain degree of heat is necessary to keep the blood fluid, and also to the action of the nerves; without either of which, no motion could be performed. Respiration or breathing is so necessary to life, that it cannot exist, even a few minutes, without the exercise of that function; and yet we shall afterwards see, that the ultimate end of respiration is to keep the body in a proper state, for the purposes of muscular motion and sensation.

The skin serves like a sheath to defend the body from injuries; the skull serves the same purpose to the brain, which is the origin of the nerves. The different membranes separate the fibres, muscles, nerves, and various organs of the body, from each other. Hence we see that there is no impropriety, in calling the human body a machine composed of bones and muscles, with their proper appendages, for the purpose of motion, at the instance of its intelligent principle.

In order to show more clearly how each part is subservient to these ends, I shall give a short account of the structure of the human body, but I must premise, that the nature of this course will prevent my entering minutely into anatomical detail. All that can be done is, to give such a general outline of anatomy and physiology, as will furnish individuals with so much knowledge of themselves, as may enable them to guard against habitual sickness.

Among the solid parts of the human frame the bones stand conspicuous. Their use is, to give firmness and shape to the body. Some of them likewise serve as armour, or defense, to guard important parts; thus the skull is admirably contrived to defend the brain; and the spine or backbone is designed, not only to strengthen the body, but to shield that continuation of the brain, called the spinal marrow, from whence originate great numbers of nerves, which pass through convenient openings of this bone, and are distributed to various parts of the body. In the structure of this, as well as every other part, the wisdom of the Creator is manifest. Had it been a single bone, the loins must have been inflexible; to avoid which, it consists of a number of small bones, articulated or joined together with great exactness, which are strengthened by compact ligaments. Hence it becomes capable of various inflections, without injuring the nerves, or diminishing that strength which is so much required.

The whole system of bones, or skeleton, is constructed of several parts, of different shapes and sizes, joining with one another in various manners, and so knit together, as best to answer to the motions which the occasions of the animal may require.

These bones serve as levers for the muscles to act on; which last serve as mechanical powers, to give the machine various motions, at the command of the will.

The muscles are fleshy fibres, attached by their extremities to the bones. When the fibres shorten themselves, the two parts into which the muscle is inserted are brought nearer; and, by this simple contrivance, all the motions of animals are performed, and their bodies carried from one place to another.

Joints are provided with muscles for performing the motions for which they are adapted; every muscle pulling the bone, to which it is attached, in its own particular direction. Hence the muscles may be considered as so many moving forces, as was before hinted; and their strength, the distance of their insertion from the centre of motion, the length of the lever to which they are attached, and the weight connected with it, determine the duration and velocity of the motions which they produce. Upon these different circumstances depend the different kinds of motion performed by various animals, such as the force of their leap, the extent of their flight, the rapidity of their course, and their address in catching their prey.

Most of the muscles act upon the bones, so as to produce the effects of a lever of the third kind, as it is termed by mechanics, where the power acts between the centre of motion and the weight; hence it has a mechanical disadvantage; as an instance of this, the muscle which bends the forearm, is inserted about one eighth or one tenth of the distance from the centre of motion that the hand is, where the weight or resistance is applied; hence the muscle must exert a force eight or ten times greater than the weight to be raised. But this disadvantage is amply compensated by making the limbs move with greater velocity; besides, if room had been given for the muscles to act with greater advantage, the limbs must have been exceedingly deformed and unwieldy. [1]

The muscles, in general, at least those which serve for voluntary motion, are balanced by antagonists, by means of which they are kept beyond their natural stretch. When one of two antagonists is contracted by the will, the other relaxes in order to give it play; or at least becomes overpowered by the contraction of the first. Also when one of such muscles happens to be paralytic, the other being no longer balanced, or kept on the stretch, immediately contracts to its natural length, and remains in that situation. The part to which it is fixed will, of course, be affected accordingly. If one of the muscles which move the mouth sideways be destroyed, the other immediately contracting, draws the mouth awry; and in that situation it remains. The same may be observed of the leg, the arm, and other parts. Some muscles assist one another in their action, while others have different actions; according to their shapes, the course of their fibres, and the structure of the parts they move.

According to the shape and nature of the bones to be moved, and of the motions to be performed, the muscles are either long, or short; slender, or bulky; straight, or round. Where a great motion is required, as in the leg, or arm, the muscles are long; where a small motion is necessary, they are short; for a strong motion they are thick, and for a weak one slender.

Some of the muscles are fastened to, and move bones; others cartilages, and others again other muscles, as may best suit the intention to be answered.

With respect to the bones, some are solid and flattened; others hollow and cylindrical. Every cylindrical bone is hollow, or has a cavity containing a great number of cells, filled with an oily marrow. Each of these cells is lined with a fine membrane, which forms the marrow. On this membrane, the blood vessels are spread, which enter the bones obliquely, and generally near their middle; from some branches of these vessels the marrow is secreted; while others enter the internal substance of the bones for their nourishment; and the reason why they enter the bones obliquely is, that they may not weaken them by dividing too many fibres in the same place.

The bones being made hollow, their strength is greatly increased without any addition to their weight; for if they had been formed of the same quantity of matter without any cavities, they would have been much weaker; their strength to resist breaking transversely being proportionate to their diameters, as is evident from mechanics.

All the bones, excepting so much of the teeth as are out of the sockets, and those parts of other bones which are covered with cartilages, are surrounded by a fine membrane, which on the skull is called pericranium, but in other parts periosteum. This membrane serves for the muscles to slide easily upon, and to hinder them from being lacerated by the hardness and roughness of the bones.

But though the apparatus which I have been describing is admirably contrived for the performance of motion; it would continue for ever inactive, if not animated by the nervous system.

The brain is the seat of the intelligent principle: from this organ, white, soft, and medullary threads, called nerves, are sent off to different parts of the body: some of them proceed immediately from the brain to their destined places, while the greater number, united together, perforate the skull, and enter the cavity of the backbone, forming what we call the spinal marrow, which may be regarded as a continuation of the brain. Portions of the spinal marrow pass through different apertures to all parts of the body.

We are not conscious of the impression of external objects on our body, unless there be a free communication of nerves, between the place where the impression is made and the brain. If a nerve be divided, or have a ligature put round it, sensation is intercepted.

There is perhaps only one sense which is common to all classes of animals, and which exists over every part of the surface of the body; I mean the sense of touch. The seat of this sense is in the extremities of the nerves distributed over the skin; and by means of it we ascertain the resistance of bodies, their figure, and their temperature.

The other senses have been thought to be only more refined modifications of the sense of touch; and the organs of each are placed near the brain on the external surface of the head. The sense of sight, for instance, is seated in the eye; the hearing in the ear; the smell in the internal membrane of the nose; and the taste in the tongue.

The light; the pulses, or vibrations of the air; the effluvia floating in the atmosphere; saline particles, or particles which are soluble in water or saliva, are the substances which act upon these four senses; and the organs which transmit their action to the nerves, are admirably adapted to the respective nature of each. The eye presents to the light a succession of transparent lenses to refract its rays; the ear opposes to the air membranes, fluids, and bones, well fitted to transmit its vibrations; the nostrils, while they afford a passage to the air in its way to the lungs, intercept any odorous particles which it contains, and the tongue is provided with spongy papillae to imbibe the sapid liquors which are the objects of taste.

It is by these organs that we become acquainted with what passes around us; by these we know that a material world exists. We may however observe, that the nervous system, besides making us acquainted with external things, gives us notice of many changes that take place within our own body. Internal pain warns us of the presence of disease; and the disagreeable sensations of hunger, thirst, and fatigue, are signs of the body standing in need of refreshment or repose.

Concerning the manner in which we become acquainted with external things, by means of the senses, we know nothing. Many hypotheses have been offered to explain this: none of them however are the result of experiment and observation. Many philosophers have supposed the universe to be filled with an extremely subtile fluid, which they have termed ethereal; and this hypothesis has been sanctioned by the illustrious authority of Newton. He however merely offered it in the modest form of a query, for the attention of other philosophers; little thinking that it would be made use of to explain phenomena which they did not understand. His query about a subtile elastic fluid pervading the universe, and giving motion and activity to inert masses of matter, and thereby causing the phenomena of attraction, gravitation, and many other appearances in nature, was immediately laid hold of by his followers, as a fact sufficiently supported, because it seemed to have the sanction of so great an authority.

This hypothesis was made use of to explain a great number of phenomena, and the physiologists, whose theories were generally influenced by the prevailing philosophy, eagerly laid hold of it to explain the phenomena of sensation, and muscular motion. When an impression was made upon any part of the external surface of the body, whether it was occasioned by heat, or mechanical impulse, they supposed, that the ether in the extremities of the nerves was set in motion. This motion, from the energy of the ether, is communicated along the nerves to the brain, and there produces such a change as occasions a consciousness of the original impression, and a reference in the mind to the place where it was made. Next they supposed, that the action of the will caused a motion of the ether to be instantly propagated along the nerves that terminate in the fibres of the muscles, which stimulated them to contraction.

Other philosophers imagined, that a tremulous motion was excited in the nerves themselves, by the action of external impulses, like the motions excited in the string of a harp. These motions they supposed to be propagated along the nerves of sense, to the brain, and from thence along the motory nerves, to the muscles.

Before they attempted this explanation of the phenomena, they should have proved the existence of such a fluid, or at least brought forward such circumstances, as rendered its existence credible. But supposing we grant them the hypothesis, it will, in my opinion, not avail much; for it is not easy to conceive how the motion of a subtile fluid, or the vibration of a nerve, can cause sensation.

Nor are the internal senses, as they are generally called, namely, memory, and imagination, any better explained on this supposition; for we cannot conceive how this nervous fluid is stored up and propelled by the will.

After all, I think we must confess, that this subject is still enveloped in obscurity. One observation is worth making, namely, that our sensations have not the smallest resemblance to the substance or impression, which causes them; thus the sensation occasioned by the smell of camphor, possesses not the smallest resemblance to small particles of camphor floating in the atmosphere; a sensation of pain has no similitude whatever to the point of a sword which occasions it; nor has the sensation of sound any resemblance to waves or tremors in the air. In our present state of knowledge, therefore, all that we can say, is, that nature has so formed us, that when an impression is made on any of the organs of sense, it causes a sensation, which forces us to believe in the existence of an external object, though we cannot see any connexion between the sensation and the object which produces it.

But though philosophers were certainly blameable for assuming an unknown cause, to account for various phenomena, yet later experiments would seem to prove that even the conjectures of Newton were not founded on slight grounds. His idea that the diamond was inflammable, has been confirmed by various experiments: and that there exists in nature a subtile fluid, capable of pervading with ease the densest bodies, the phenomena of electricity would seem to show, and some late experiments render it by no means improbable, that this fluid or influence, acts a conspicuous part in the nervous system. That it exists in great quantity in animal bodies, is evident on gently rubbing the back of a cat in the dark; and it would seem that, in some instances, it may be given out or secreted by the nerves. We shall afterwards see that the seat of vision is a delicate expansion of a large nerve which comes from the brain, and is spread out on the bottom of the eye; and flashes of light, or a kind of sparkling, is often seen to dart from the eyes of persons in high health, and possessed of much nervous energy. These luminous flashes are very apparent in the dark in some animals; such as the lion, the lynx, and the cat; and it is difficult to account for this appearance unless we suppose it electrical.

The experiments of Galvani and others, have however proved beyond all doubt, that this fluid, when applied to the nerves and muscles, is capable of exciting various sensations and motions. To produce this fluid by the application of two metals, it is necessary that one of them should be in such a situation, as to be easily oxidable, while the other is prevented from oxidation. If a piece of zinc be put into water, no change will take place; but if a piece of silver be put along with it, the zinc will immediately oxidate, by decomposing the water, and a current of electricity will pass through the silver. If the upper and under surfaces of the tongue be coated with two different metals, one of which is easily oxidable, and these be brought into contact, a sensation is produced resembling taste, which takes place suddenly, like a slight electrical shock. This taste may likewise be produced by applying one part of the metals to the tongue and the other to any part of the body deprived of the cuticle, and bringing them in contact.

The sensation of light may be produced in various ways; such as by applying one metal between the gum and the upper lip, and the other under the tongue; or by putting a silver probe up one of the nostrils, and a piece of zinc upon the tongue; a sensation resembling a very strong flash of light is perceived in the corresponding eye, at the instant of contact.

But the experiments which tend most strongly to prove what I have hinted, are made in the following manner. Lay bare a portion of a great nerve leading to any muscle or limb of an animal, for instance, the leg of a frog separated from the body. Touch the bared nerve with a piece of zinc, and the muscle with a piece of silver, and strong contractions take place the instant these metals are brought into contact. The same effect may be produced by placing a piece of silver on a larger piece of zinc, and putting a worm or a leech on the silver; in moving about, the instant it touches the zinc it is thrown into strong convulsions.

These phenomena have been clearly proved to be electrical; for by a number of pieces of these metals, properly disposed, strong shocks can be given, the electrometer can be affected, a Leyden vial charged, the electric spark seen, and combustible bodies inflamed.

Some animals likewise possess the power of accumulating this influence in a great degree; for instanc the torpedo, and electrical eel, which will both give strong shocks; and if the circuit have a small interruption a spark may be seen, as was shown by Mr. Walsh. On dissecting these fish, Mr. Hunter found an organ very similar to the pile of Volta; it consists of numerous membranaceous columns, filled with plates or pellicles, in the form of thin disks, separated from each other by small intervals, which intervals contain a fluid substance; this organ, like the pile of Volta, is capable of giving repeated shocks, even when surrounded by water.

It is not absolutely necessary to use two metals to produce the galvanic phenomena; for if one side of a metal be made to oxidate, while the other is prevented from oxidation, these appearances will still be produced. It is not indeed necessary to use any metal; for a piece of charcoal, oxidated in the same way, produces galvanism; so does fresh muscular fibre, and perhaps any substance capable of oxidation. The most striking circumstance in galvanism, is, that it accompanies oxidation, and is perhaps never produced without it. But oxidation is always going on in the body by means of respiration and the circulation of the blood. We shall afterwards find reason to believe, that the oxygen, received from the atmosphere by the lungs, is the cause of animal heat, and probably of animal irritability; and it is perhaps not unreasonable to suppose, that the nervous influence or electricity may be separated by the brain, and sent along the nerves, which seem the most powerful conductors of it, to stimulate the muscles to action.

What the nature of the electric fluid is, we are ignorant; some galvanic experiments have led me to suppose that it may be hydrogen, which when combined with caloric appears in the form of gas, but when pure, or perhaps in a different state, may be capable of passing through solid bodies in the form of electricity.

Having given this short view of the human body, considered as a machine composed of bones, muscles, and nerves, I shall proceed to state the different subjects which I shall consider in this course. It is extremely difficult to begin a course like this; for we must either enter abruptly into the middle, or the outset must be in some measure tedious and dry. I have chosen however rather to hazard the latter appellation, with respect to this lecture, than to enter more abruptly into the subject, in order to make it more entertaining. As we proceed, I trust you will feel an increasing interest in the subject; and, I think I may venture to promise, that this will be found the least entertaining lecture in the course. The subjects will be illustrated by experiments, in order to render the deductions more striking.

I shall next proceed to consider the phenomena of respiration, and animal heat; after which I shall explain the circulation of the blood; and the phenomena of digestion and nutrition. I shall then examine, more minutely than has been done in this lecture, the connexion of man with the external world, which will lead to a discussion of the different senses; vision, hearing, smelling, tasting, and feeling.

When these subjects have been discussed as fully as our time will allow, I shall examine, at considerable length, the manner in which the powers that support life, which have been improperly called by physiologists, the nonnaturals, act upon the body This will naturally lead to a fuller explanation of the system which I have attempted to defend, in my lecture on health. And, after I have fully explained the laws by which the irritable principle is regulated, I shall proceed to show, how those variations from the healthy state, called diseases, are produced; I shall point out the difference that exists between the debility which is brought on by the diminished action of the powers which support life, and that which results from their too powerful action; I shall then inquire into the nature of diseases of increased excitement; and after having shown how the undue action of the powers which support life, operates in producing disease, I shall endeavour to lay down such rules for the preservation of health, as are the result of reasoning on these subjects, and are also confirmed by experience.

[1] [FIGURE] Suppose AC to be a lever, held in equilibrio by the force B and weight W, then the whole momentum exerted at B must be equal to that at W, but the forces will be different. For B x AC = W x AB, and if AC = 10AB, then a force equal to ten times the weight to be raised must be exerted by the muscle. Hence we see, that in the actions of muscles there is a loss of power, from their insertions being nearer the fulcrum than the weight. For example, suppose the deltoid muscle to act and raise a weight of 55 lb.: the weight of the arm is 5 lb., and the distance of its insertion is only 1/3 of the arms length, hence the force exerted must be (55 + 5) x 3 = 180 lb.

[FIGURE] But by this contrivance we gain a greater extent of motion, and also a greater velocity, and both with less contraction. Let A be the centre of motion, or articulation; B the insertion of a muscle, and AC the length of the lever or bone; then, by a contraction only equal to B_b_, C is carried through C_c_, which is to B_b_ as AC to AB. It is obvious also, that the velocity is greater, since C moves to *c* in the same time as B to *b*.

A loss of power is likewise occasioned by the obliquity of the muscular action, and the oblique direction of the fibres.

For, in this case, there is a compound of two forces, and a consequent loss of power: for the forces are proportioned to the two sides of a parallelogram, but the effects produced are proportioned only to the diagonal.

LECTURE II. RESPIRATION.

In the last lecture I took a short view of the human body, as a moving machine, regulated by the will. We shall now proceed to examine some of its functions more particularly.

I need not tell any of my audience, how necessary air is to the living body; for every person knows that we cannot live when excluded from this fluid; but, before we can understand the manner in which it acts on the body, we must become acquainted with some of its properties.

That the air is a fluid, consisting of such particles as have little or no cohesion, and which slide easily among each other, and yield to the slightest force, is evident from the ease with which animals breathe it, and move through it. Indeed from its being transparent, and therefore invisible, as well as from its extreme tenuity, and the ease with which bodies move through it, people will scarcely believe that they are living at the bottom of an aerial ocean, like fishes at the bottom of the sea. We become, however, very sensible of it, when it flows rapidly in streams or currents, so as to form what is called a wind, which will sometimes act so violently as to tear up the strongest trees by the roots, and blow down to the ground the best and firmest buildings.

Some may still be inclined to ask, what is this air in which we are said to live? We see nothing; we feel nothing; we find ourselves at liberty to move about in any direction, without any hindrance. Whence then comes the assertion, that we are surrounded by a fluid, called air? When we pour water out of a vessel, it appears to be empty; for our senses do not inform us that any thing occupies the place of the water, for instance, when we pour water out of a vial. But this operation is exactly similar to pouring out mercury from a vial in a jar of water, the water gets in and supplies the place of the mercury; so does the air which supplies the place of the water; and this air will prevent water from rising, or filling a vessel which contains it.

Hence we see that air possesses similar appearances of impenetrability with other matter: for it excludes bodies from the space which itself occupies.

Air being therefore material must have weight; and we shall accordingly find, that a quart of it weighs about fifteen grains. But a quart of water weighs about two pounds; this fluid therefore is nearly a thousand times heavier than air.

But though air is so much lighter than water, yet, because it extends to a considerable height above the surface of the earth, it is evident, that it must press strongly on the surfaces of bodies. It is thought to extend nearly fifty miles above the surface of the earth, and must therefore press heavily on this surface. This may be evinced by different experiments, performed by means of the air pump.

Another property of the air, by which it is distinguished from most other fluids, is its elasticity. It may be compressed into a less space than it naturally occupies, and when the compressing force is removed, it expands to its former bulk, by its spring or elasticity. Indeed it is always compressed into less space than it would naturally occupy, by the weight of the superincumbent air.

The trachea, or windpipe, commences at the further end of the mouth, between the root of the tongue, and the passage into the stomach: its upper part is termed the larynx; it forms the projection in the fore part of the neck, which is more prominent in the male than the female: its opening is called the glottis, and is covered with a small valve, or lid, called the epiglottis, which is open while we breathe, but shuts when we swallow any thing, to prevent its getting into the lungs: sometimes, however, particularly when we attempt to speak at the time we swallow, a small portion of our food or drink gets into the larynx, and excites violent coughing until it is thrown back again.

The windpipe is composed of cartilaginous rings, covered with membrane, which keep it open: after having run downwards for the space of a few inches, it divides into two great branches, each of which is subdivided into a vast number of ramifications, ultimately terminating in little vesicles, which, when distended with air, make up the greatest part of the bulk of the lungs.

The cavity in which the lungs are contained is called the thorax, or chest: and is bounded by the ribs, and backbone or spine, and separated from the abdomen by a muscular membrane, called the diaphragm. The thorax, by the action of the diaphragm and intercostal muscles, is alternately enlarged and diminished. Suppose then the thorax to be in its least state; if it become larger, a vacuum will be formed, into which the external air will descend by its weight, filling and distending the vesicles of the lungs.

The thorax, thus dilated, is brought back to its former magnitude, principally by the relaxation of the muscles, which distended it, and the natural elasticity of the parts, aided by the contraction of the abdominal muscles; the thorax being thus diminished, a quantity of air is expelled from the lungs. The muscles which distend the thorax beginning again to act, the air reenters; and this alternate dilatation and contraction, is called respiration. The entrance of the air into the lungs, is termed inspiration, and its expulsion, expiration.

To form a more accurate idea of the manner in which respiration is performed, let us suppose this room to be filled with water. On enlarging the thorax, in the manner before mentioned, the water by its weight would rush in, and fill the newly formed void; and, upon the diminution of the capacity of the thorax, a part of this water would be expelled. Just in the same manner the air will alternately enter and be expelled from the lungs by this alternate dilatation and contraction of the thorax.

Respiration is a function of such consequence, that death follows if it is suspended for a few minutes only. By means of this function the blood is elaborated, and rendered fit to nourish the body; by means of it the system is, most probably, supplied with irritability; by means of it the nervous energy is, most likely, conveyed into the body, to be expended in sensation, and muscular motion. It appears, likewise, that in this way, animals are supplied with that heat which preserves their temperatures nearly the same, whatever may be the temperatures of surrounding bodies.

If any number of inanimate bodies, possessed of different degrees of heat, be placed near each other, the heat will begin to pass from the hotter bodies to the colder, till there be an equilibrium of temperature. But this is by no means the case with respect to animated matter; for whatever be the degree of heat peculiar to individual animals, they preserve it, nearly unchanged, in every temperature, provided the temperature be not altogether incompatible with life or health. Thus, we find, from experiments that have been made, that the human body is not only capable of supporting, in certain circumstances, without any material change in its temperature, a degree of heat considerably above that at which water boils; but it likewise maintains its usual temperature, whilst the surrounding medium is several degrees below frost.

It is evident, therefore, that animals neither receive their heat from the bodies which surround them, nor suffer, from the influence of external circumstances, any material alterations in that heat which is peculiar to their nature. These general facts are confirmed and elucidated by many accurate and well authenticated observations, which show, that the degree of heat in the same genus and species of the more perfect animals, continues uniformly the same, whether they be surrounded by mountains of snow, in the neighbourhood of the pole, or exposed to a vertical sun, in the sultry regions of the torrid zone.

This stability and uniformity of animal heat, under such a disparity of external circumstances, and so vast a latitude in the temperature of the ambient air, prove, beyond doubt, that the living body is furnished with a peculiar mechanism, or power of generating, supporting, and regulating its own temperature; and that this is so wisely adapted to the circumstances of its economy, or so dependent upon them, that, whatever be the temperature of the atmosphere, it will have very little influence either in diminishing or increasing that of the animal.

In order that we may see how this effect is produced, we must examine the chemical properties of the air. Previously to this, however, it will be necessary to point out briefly how bodies are affected, with respect to heat, when they change their form.

When a body passes from a state of solidity to that of fluidity, it absorbs a quantity of heat, which becomes chemically combined with it, and insensible to the touch or the thermometer; in the same manner, when it passes from a fluid state to that of vapour or gas, it combines with a still larger quantity of heat, which remains latent in it, so long as it continues in the state of gas, but when it returns to the liquid or solid state, it gives out the heat which was combined with it, which, being set at liberty, flows into the surrounding bodies, and augments their temperature.

This is evinced by the conversion of ice into water, and of water into steam; and by the return of steam into water. It is evinced likewise by the evaporation of ether, and by numberless other experiments.

Modern chemistry has shown that the atmosphere is not a homogeneous fluid, but consists of two elastic fluids, endowed with opposite and different properties.

If a combustible body, for instance a candle, be confined in a given quantity of atmospheric air, it will burn only for a certain time; after it is extinguished, if another combustible body be lighted and immersed in the same air, it will not burn, but will immediately be extinguished.

It has been proved by chemical experiments, that in this instance, the combustible body absorbs that portion of the air which is fitted for combustion, but produces no change on that which is unfit: so that, according to this, the air of the atmosphere consists of two elastic fluids, one of which is capable of supporting combustion, and the other not; and that they exist in the proportion of one part of the former to three of the latter nearly.

These two parts may be separated from each other, and experiments made with them.

Many metals, and particularly manganese, when exposed to the atmosphere, attract the combustible air from it, without touching the other; and it may be procured from these metals by the application of heat, in very great purity.

Because this air is essential to the formation of acids, it has been called by chemists the acidifying principle, or oxygen gas.

On plunging a combustible body into the remaining air, it is instantly extinguished; an animal in the same situation is immediately deprived of life: from this latter circumstance this air has been called azote, or azotic gas. If we take three parts of azote and one of oxygen, and mix them together, we shall form an air in every respect similar to that of the atmosphere.

If I plunge a piece of iron, previously heated, into oxygen gas, it will burn with great brilliancy, the gas will be diminished in quantity, and the iron augmented in weight, and this increase of weight in the metal will be in proportion to the oxygen which has disappeared: at the same time a great quantity of heat is given out. This is the heat which was combined with the oxygen in the state of gas, and which now becomes free, when the oxygen becomes solid and joins with the iron.

The same phenomena take place when phosphorus is burned in oxygen gas; the gas becomes diminished, the phosphorus increased, in weight, and converted into an acid, and a great quantity of heat is given out. The same is the case when charcoal is burned in this gas. In short, in every instance of combustion, the oxygen combines with the combustible body, and at the same time gives out its heat, which supported it in the form of gas. This is the case of the combustion of coal in a common fire, as well as in other cases of combustion; the heat comes from the air, and not from the coal.

When we examine the phenomena of respiration with attention, we shall find them very analogous to those of combustion. A candle will not burn in an exhausted receiver: an animal in the same situation ceases to live.

When a candle is confined in a given quantity of atmospheric air, it will burn only for a certain length of time. On examining the air in which it has been burned, the oxygen is found to be all extracted, nothing remaining but azotic gas, and a quantity of carbonic acid gas, produced by the union of the charcoal of the candle with the oxygen of the atmospheric air.

In the same manner, if an animal be confined in a given quantity of atmospheric air, it will live only a short time; on examining the air in which it has ceased to live, it will be found to have lost its oxygen: what remains being a mixture of azotic and carbonic acid gases.

When a candle is enclosed in a given quantity of pure oxygen gas, it will burn four times as long as in the same quantity of atmospheric air.

In the same manner it has been proved, that an animal will be four times as long in consuming a given quantity of pure oxygen gas, as in rendering unfit for respiration the same quantity of atmospheric air.

Here then we observe a striking similarity between combustion and animal respiration. The ancients seem to have had a more accurate idea of respiration than most of the philosophers who followed them. They supposed that the air contained a principle proper for the support and nourishment of life, which they called pabulum vitae. This idea, which was unconnected with any hypothesis, was followed by systems destitute of foundation. Sometimes it was thought that the air in the lungs incessantly acted as a stimulus or spur to drive on the circulation; sometimes the lungs were considered in the light of a pair of bellows, or fan, to cool the body, which was supposed to be heated by a thousand imaginary causes: and when philosophers were convinced, by experiments, that the bulk of the air was diminished by respiration, they explained it by saying, that the air had lost its spring.

Modern chemistry however enables us to explain the phenomena of respiration in a satisfactory manner.

In order to see this, we shall proceed to examine the changes produced by respiration; firstly, on the air, and secondly, on the blood.

The air which has served for respiration, is found to contain a mixture of azotic and carbonic acid gas, with a small quantity of oxygen gas; and a considerable quantity of water is thrown off from the lungs, in the form of vapour, during respiration.

From a variety of facts, it appears that oxygen gas is decomposed in the lungs during respiration; a part of it unites, as we shall afterwards see, with the iron contained in the blood, and converts it into an oxid; another and greater portion unites with the carbon, brought by the venous blood from all parts of the body to the lungs, and thus forms carbonic acid gas; while another portion of the oxygen unites with the hydrogen, brought in the same manner by the blood, and forms water. Thus then we are able to account for the different products of respiration.

Hence we see, that the explanation of animal heat follows as a simple and beautiful corollary from the theory of combustion; and we may consider respiration as an operation in which oxygen gas is continually passing from the gaseous to the concrete state; it will therefore give out at every instant the heat which it held in combination, and this heat, being conveyed by the circulation of the blood to all parts of the body, is a constant source of heat to the animal.

These facts likewise enable us to explain the reason, why an animal preserves the same temperature, notwithstanding the various changes which occur in the temperature of the surrounding atmosphere. In winter the air is condensed by the cold, the lungs therefore receive a greater quantity of oxygen in the same bulk, and the heat extricated will be proportionally increased. In summer, on the contrary, the air being rarefied by the heat, a less quantity of oxygen will be received by the lungs during each inspiration, and consequently the heat which is extricated must be less.

For the same reason, in northern latitudes, the heat extricated by respiration will be much greater than in the southern. By this simple and beautiful contrivance, nature has moderated the extremes of climate, and enabled the human body to bear vicissitudes which would otherwise destroy it.

Of all the phenomena of the animal body, there is none at first sight more remarkable, than that which animals possess of resisting the extremes of temperature.

The heat of the body, as has already been observed, continues at the same degree, though the temperature of the atmosphere be sometimes considerably hotter, at other times considerably colder, than the animal body: so that man is able to live, and to preserve the temperature of health, on the burning sands of Africa, and on the frozen plains of Siberia.

The alterations of temperature which the human body has been known to bear, without any fatal or even bad effects, are not less than 400 degrees or 500 degrees of Fahrenheit. The natural heat of the human body is 96 degrees or 97 degrees. In the West Indies, the heat of the atmosphere is often 98 degrees or 99 degrees, and sometimes rises even to 126 degrees, or 30 degrees above the temperature of the human body, notwithstanding which, a thermometer put in the mouth points to 96 degrees or 97 degrees. The inhabitants of the hot regions of Surinam support, without inconvenience, the heat of their climate. We are assured that in Senegal, about the latitude of 17 degrees, the thermometer in the shade generally stands at 108 degrees, without any fatal effects to men or animals. The Russians often live in places heated by stoves to 108 degrees or 109 degrees, and some philosophers in this country, by way of experiment, remained a considerable time in a room heated above the boiling point of water.

On the other hand, an equal excess of cold seems to have no greater effect in altering the degree of heat proper to animal bodies. Delisle has observed a cold in Siberia 70 degrees below the zero of Fahrenheit's scale, notwithstanding which animals lived. Gmelin has seen the inhabitants of Jeniseisk under the 58th degree of northern latitude, sustaining a degree of cold, which in January became so severe, that the spirit in the thermometer was 126 degrees below the freezing point. Professor Pallas in Siberia, and our countrymen at Hudson's Bay, have experienced a degree of cold almost equal to this. All these facts tend to prove, that the heat of animals continues without alteration in very different temperatures. Hence it is evident that they must be able to produce a greater degree of heat, when surrounded by a cold medium; and on the contrary, that they must effect a diminution of the heat, when the surrounding medium is very hot.

All these circumstances may be accounted for, by the principles we have laid down; the decomposition of oxygen in the lungs.

There have not been wanting, however, some very eminent physiologists, who have contended that animal heat is produced chiefly by the nerves. They have brought forward in proof of this the well known fact, that when the spinal marrow is injured, the temperature of the body generally becomes diminished; and that in a paralytic limb the heat is less than ordinary, though the strength and velocity of the pulse remain the same. These facts, and others of a similar nature, have induced them to believe, that the nervous system is the chief cause and essential organ of heat; and they have adduced similar arguments, to prove that nutrition is performed by the nerves, for a limb which is paralytic from an injury of the nerves, wastes, though the circulation continues. The truth is, that the nerves exert their influence upon these, and all other functions of the body, and modify their action. The liver secretes bile, but if the nerves leading to it be destroyed, the secretion of bile will cease; but who will say, that the bile is secreted by the nerves? The nitric acid will dissolve metals, and this solution will go on more quickly if heat be applied; but surely the nitric acid is the solvent, the heat being only an aiding cause.

But though the human body has been so wisely constructed, as to bear, without inconvenience, a considerable variation of temperature; yet this latitude has its limits, which depend upon the capability of extricating heat from the atmosphere. There must be a limit below which the diminution of heat takes place faster than its production. If this be continued, or increased, the heat of the animal must diminish, the functions lose their energy, and an insuperable inclination to sleep is felt, in which if the sufferer indulge, he will be sure to wake no more.

This is confirmed by what happened to Sir Joseph Banks and his party on the heights of Terra del Fuego. Dr. Solander, who had more than once crossed the mountains which divide Sweden from Norway, well knew that extreme cold produces an irresistible torpor and sleepiness, he therefore conjured the company to keep always in motion, whatever exertion it might require, and however great might be their inclination to rest. Whoever sits down, says he, will sleep; and whoever sleeps will wake no more. Thus, at once admonished and alarmed, they set forward; but, while they were still upon the naked rocks, the cold was so intense, as to produce the effects which had been so much dreaded. Dr. Solander himself was the first who found the inclination against which he had warned others, irresistible; and insisted on being suffered to lie down. Sir Joseph entreated and remonstrated in vain; he lay down upon the ground, though it was covered with snow; and it was with great difficulty that his friend kept him from sleeping. One of his black servants also began to linger, having suffered from the cold in the same manner as the Doctor. Partly by persuasion, and partly by force, they were got forwards; soon however they both declared that they would go no further. Sir Joseph had recourse again to entreaty and expostulation, but these produced no effect: when the black was told, that if he did not go on, he would shortly be frozen to death; he answered, that he desired nothing so much as to lie down and die. The Doctor did not so explicitly renounce his life, but said, he would go on, if they would first allow him to take some sleep, though he had before told them, that to sleep was to perish. They both in a few minutes fell into a profound sleep, and after five minutes Sir Joseph Banks happily succeeded in waking Dr. Solander, who had almost lost the use of his limbs; the muscles were so shrunk, that his shoes fell from his feet; but every attempt to recal the unfortunate black to life proved unsuccessful.

As the circulation of the blood is the means by which the heat produced is conveyed to all parts of the body; and as it is a function of the highest importance, I shall, in the next lecture, proceed to the consideration of it.

LECTURE III. CIRCULATION OF THE BLOOD.

Two kinds of motion may be distinguished in the animal economy; the one voluntary, or under the command of the will, which takes place at certain intervals, but may be stopped at pleasure. The other kind of motion is called involuntary, as not depending on the will, but going on constantly, without interruption, both when we sleep and when we wake.

Of the first kind is the motion of the limbs, of which I have already spoken in general terms; the object of which is, to change the situation of the animal, and carry it where the will directs.

Among the involuntary motions, the most remarkable is the circulation of the blood, which I shall proceed to consider in this lecture.

There is one motion, however, which claims a middle place between the voluntary and involuntary; I mean respiration. This action is so far under the command of the will, that it may be suspended, increased, or diminished in strength and frequency: but we can only suspend it for a very short time; and it goes on regularly during sleep, and in general, even when we are awake, without the intervention of the will; its continuation being always necessary, as we have already seen, to support life.

The motion of the fluids in the living body is regulated by very different laws, from those which govern the motion of ordinary fluids, that depend upon their gravity and fluidity: these last have a general centre of gravitation to which they incessantly tend. Their motion is from above downwards, when not prevented by any obstacles; and when they meet with obstruction, they either stop till the obstacle is removed, or escape where they find the least resistance. When they have reached the lowest situations, they remain at rest, unless acted upon by some internal impulse, which again puts them in motion.

But the motion of the fluids in an animal body, is less uniform, constant, and regular; it takes place upwards as well as downwards, and overcomes numerous obstacles; it carries the blood from the interior parts of the body to the surface, and from the surface back again to the internal parts; it forces it from the left side of the body to the right, and with such rapidity that not a particle of the fluid remains an instant in the same place.

The principal organ concerned in the circulation of the blood, is the heart; which is a hollow muscle, of a conical figure, with two cavities, called ventricles; this organ is situated in the thorax or chest; its apex or point is inclined downwards and to the left side, where it is received in a cavity of the left lobe of the lungs.

At the basis of the heart on each side are situated two cavities, called auricles, to receive the blood; and these contracting, force the blood into the ventricles, which are two cavities in the heart, separated from each other by a strong muscular partition. The cavity which is situated on the right side of the heart, is called the right ventricle, and that on the left the left ventricle. From the right ventricle of the heart issues a large artery, called the pulmonary artery, which goes to the lungs, and is there divided and subdivided into a vast number of branches, the extremities of which are too small to be visible. These ultimate ramifications unite again into larger branches; these again into branches still larger, and so continually, till at last they form four tubes, called the pulmonary veins, which are inserted into the left auricle of the heart,

From the left ventricle of the heart there issues another large artery, called the aorta, which, in its passage, sends off branches to the heart, arms, legs, head, and every other part of the body. These branches, in the course of their progress, are divided and subdivided into innumerable minute ramifications, the last of which are invisible. These small ramifications unite again into branches continually larger and larger, till they form two great tubes, called the venae cavae; which large veins are inserted into the right auricle of the heart; where a vein, termed the coronary vein of the heart, which returns the blood from the heart itself, also terminates.

From what has been said, it will be evident, that strictly speaking, there are only two arteries and seven veins in the body; one pulmonary artery, which carries the blood from the right ventricle of the heart to the lungs, and four pulmonary veins, which bring it back again; then the aorta or large artery, which carries the blood from the left ventricle of the heart to all parts of the body; the two venae cavae, and the coronary vein of the heart, which bring it back again.

At the beginning of both arteries, where they leave the heart, are placed valves, which allow the blood to flow freely from the heart into the arteries, but which prevent its return to the heart. There are likewise valves between the auricles and ventricles, which permit the blood to flow from the former into the latter, but prevent its return into the auricles. The veins are likewise furnished with valves, which allow the blood to flow from their minute branches along the larger toward the heart, but prevent its returning to these minute branches.

The blood being brought back from all parts of the body into the right auricle of the heart, distends this cavity, and thus causes it to contract; this auricle, by contracting, forces the blood into the right ventricle; this muscular cavity being distended and irritated by the blood, contracts, and propels the blood through the pulmonary artery into the lungs: from hence it is brought back by the pulmonary veins, to the left auricle of the heart, by whose contraction it is forced into the left ventricle. The contraction of this ventricle propels the blood, with great force, into the aorta, through the innumerable ramifications of which, it is carried to every part of the body, and brought back by veins, which accompany these arterial ramifications, and form the venae cavae, which conduct the blood into the right auricle of the heart, from whence it is again sent into the right ventricle, which sends it through the pulmonary artery, to the lungs; the pulmonary veins bring it back again to the heart, from whence it is propelled through the aorta, to all parts of the body: thus running a perpetual round, called the circulation of the blood.

Thus then we see, that the circulation consists of two circles or stages, one through the lungs, which may be called the pulmonary, or lesser circle, and the other through all parts of the body, which may be termed the aortal, or greater circle.

That the blood circulates in this manner, is evident, from the valves placed at the origin of the arteries, and in the large branches of the veins, which prevent the return of the blood to the heart, in any other manner than that I have described. This is likewise evident, in the common operation of blood letting: when the arm is tied, the vein swells below the ligature, instead of above, and we do not make the opening above the ligature, or on the side next the heart. If the vein were opened above the ligature, it would not bleed. For it only swells next the hand, which shows that the blood does not flow into the vein downwards from the heart, but upwards from the hand.

If the ligature be too tight, the blood will not flow through the opening in the vein. The reason of this, is, that the artery is compressed, in this case, as well as the vein; and as the veins derive their blood from the arteries, it follows that if the blood's motion be obstructed in the latter, none can flow from them into the former: when we wish to open an artery, the orifice must be made above the ligature.

Another proof of the circulation being performed in this manner, is derived from microscopic observations, on the transparent parts of animals, in which the blood can be seen to move towards the extremities, along the arteries, and return by the veins.

The blood, however, does not flow out of the heart into the arteries in a continued stream, but by jets, or pulses; when the ventricles are filled with blood from the auricles, this blood stimulates them, and thereby causes them to contract; by such contraction, they force the blood, which they contain, into the arteries; this contraction is called the systole of the heart. As soon as they have finished their contraction, they relax, till they are again filled with blood from the auricles, and this state of relaxation of the heart, is called the diastole.

This causes the pulsation or beating of the heart. The arteries must, of course, have a similar pulsation, the blood being driven into them only by starts; and accordingly we find it in the artery of the wrist; this beating we call the pulse; the like may also be observed in the arteries of the temples, and other parts of the body. The veins, however, have no pulsation, for the blood flowing on, in an uninterrupted course, from smaller tubes to wider, its pulse becomes entirely destroyed.

The different cavities of the heart do not contract at the same time; but the two auricles contract together, the ventricles being at that time in a state of relaxation; these ventricles then contract together, while the auricles become relaxed.

Both the arteries and veins may be compared to a tree, whose trunk is divided into large branches; these are subdivided into smaller, the smaller again into others still smaller; and we may observe, likewise, that the sum of the capacities of the branches, which arise from any trunk, is always greater than the capacity of the trunk.

The minutest branches of the arteries, being reflected, become veins, or else they enter veins that are already formed, by anastomosis, as it is called; the small veins continually receiving others, become, like a river, gradually larger, till they form the venae cavae, which conduct the blood to the heart.

Anatomical injections prove, that the last branches of the arteries terminate in the beginning of veins; but it is the opinion of many celebrated physiologists, that the arteries carry the blood to the different parts of the body to nourish them, and that the veins commence by open mouths, which absorb or suck up what is superfluous, and return it back to the heart.

From what has been said, it must be evident that there is a considerable resemblance between the circulation of the blood in the animal body, and the circulation of the aqueous fluid on the surface of the globe. In the latter case the water is raised from the ocean, by the heat of the sun, and poured down upon the dry land, in minute drops, for the nourishment and economy of its different parts. What is superfluous is collected into little rills; these meeting with others, form brooks; the union of which produce rivers, that conduct the water to its original source, from which it is again circulated.

In the same manner, the blood is sent by the heart to different parts of the body, for the nourishment and economy of its different parts; what is superfluous is brought back by veins, which, continually uniting, form those large trunks, which convey the vital fluid to the heart.

The blood does not circulate, however, in the manner which I have mentioned, in all parts of the body; for that which is carried by arteries to the viscera, serving for digestion, such as the stomach, bowels, mesentery, omentum, and spleen, is collected by small veins which unite into a large trunk called the vena portarum; this vein enters the liver, and is subdivided in it like an artery, distributing through the liver a great quantity of blood, from which the bile is secreted: and, having served this purpose, the blood is collected by small veins; these unite and form the hepatic vein, which pours the blood into the vena cava, to be conducted to the heart.

The reason of this deviation, is probably, to diminish the velocity of the blood in the liver, for the secretion of the bile; which could not have been effected by means of an artery.

The force which impels the blood, is, first, the contraction of the heart, which propels the blood into the arteries with great velocity; but this is not the only force concerned in keeping up the circulation; this is evident, from the diminished heat, and weakened pulse, in a paralytic limb, which ought not to take place, if the blood were propelled merely by the action of the heart.

The arteries are possessed of an elastic and muscular power, by means of which they contract when they are distended or stimulated. It is however by the muscular power alone, that they assist in propelling the blood; for the elasticity of their coats can serve no other purpose than preserving the mean diameter of the vessel. If we suppose the arteries to be dilated by the blood, poured into them by the heart, they will, by their contraction, as elastic tubes, undoubtedly propel the blood: but supposing them to be perfectly elastic, the force of the heart will be just as much diminished in dilating them as the force of the blood is increased by their contraction. We are not however acquainted with any substance perfectly elastic, or which restores itself with a force equal to that with which it was distended: hence the elastic power of the arteries will subtract from, instead of adding to, the power of the heart. It is evident, therefore, that it must be by the muscular power of the arteries, which causes them to contract like the heart, that they propel the blood.

That such is the case, appears from the muscular structure of the arteries observed by anatomists; as also from the effects of mechanical irritation of their coats, which causes them to contract; this is likewise evident from the inflammation produced by the application of stimulating substances to particular parts; for instance, cantharides and mustard. It appears likewise, from the secretion in some parts being preternaturally increased, while the motion of the general mass of the blood continues unaltered.

The contraction of the arteries always propels the blood towards the extreme parts of the body: this must necessarily happen, because the valves at the origin of the arteries prevent its return to the heart, it must therefore move in the direction in which it finds least resistance.

If it were not for this muscular power of the arteries, the force of the heart would not alone be able to propel the blood to the extreme parts of the body, and overcome the different kinds of resistance it has to encounter. Among the causes that lessen the velocity of the blood, may be mentioned the increasing area of the artery; for it was before observed, that the sum of the cavities of the branches from any trunk exceeded the cavity of the trunk: and from the principles of hydrostatics, the velocities of fluids, propelled by the same force, in tubes of different diameters, are inversely as the squares of the diameters, so that in a tube of double the diameter, the velocity will only be one fourth; in one of the triple, only one ninth: and since the arteries may be looked upon as conical, it is evident that the velocity of the blood must be diminished from this cause.

The curvilinear course of the arteries likewise gives considerable resistance; for at every bending the blood loses part of its momentum against the sides; and this loss is evidently proportioned to the magnitude of the angle, at which the branch goes off. Convolutions are frequently made, in order to diminish the force of the blood in particular organs; this is especially the case with the carotid artery before it enters the brain.

The angles which the ramifications of the arteries make, are greater or more obtuse nearer the heart, and more acute as the distance increases; by which means the velocity of the blood is rendered more equal in different parts.

The anastomosing or union of different branches of arteries, likewise retards the velocity of the blood, the particles of which, from different vessels, impinging, disturb each other's motion, and produce a compound force, in which there is always a loss of velocity: and it is evident, from the composition of forces, that this loss must be proportioned to the obliquity of the angle at which the vessels unite.

The adhesion of the blood to the sides of the vessels, likewise causes a loss of velocity in the minuter branches, which may be owing to a chemical affinity: the viscidity or imperfect fluidity of the blood is another retarding cause. All these causes united, would render it impossible for the heart to propel the blood with the velocity with which it moves in the very minute branches of the arteries, if these arteries were not endowed with a living muscular power like the heart, by which they contract and propel their contents.

In the veins, the motion of the blood is occasioned partly by the vis a tergo, and partly by the contraction of the neighbouring muscles, which press upon the veins; and these veins being furnished with valves, the return of the blood towards the arteries is prevented; it must therefore move towards the heart.

That the contraction of the muscles of the body tends very much to promote the circulation of the blood, is evident, from the increase of the circulation from exercise, and likewise from the languid motion of the blood in sedentary persons, and those given to indolence. Hence we may account for the different diseases to which such persons are subject, and know how to apply the proper remedies. Hence likewise, we see the reason why rest is so absolutely necessary in acute and inflammatory diseases, where the momentum of the blood is already too great.

It has been doubted by anatomists, whether the veins were possessed with muscular power; but this seems now to be confirmed. Haller found the vena cava near the heart to contract on the application of stimulants, though he could see no muscular fibres; these, however, have been discovered by succeeding anatomists.

The magnitude of the veins is always greater than that of the corresponding arteries; hence the velocity of the blood must be less in the veins; and hence likewise we may account for their want of pulsation; for the action of the heart upon the arteries is at first very great; but as we recede from the heart, this effect becomes less perceptible; the arterial tube increases both in size and muscularity, in proportion to its distance from the source of circulation. The powers of the heart are spent in overcoming the different resistances which I have noticed, before the blood enters the veins; hence the blood will flow uniformly in these last.

The blood is subject in the veins to retarding causes, similar to those which operate in the arteries, but perhaps not in an equal degree; for the flexures are less frequent in the veins than in the arteries. As the capacity of the arterial tube increases with its distance from the heart, the velocity, from this cause, as has already been observed, is continually diminished; but a contrary effect takes place in the veins; for the different branches uniting, form trunks, whose capacities are smaller than the sums of the capacities of the branches, hence the velocity of the blood in the veins will increase as it approaches the heart.

Another retarding cause may be mentioned, namely, gravity, which acts more on the venous than the arterial system. The effects of gravity on the veins may be exemplified, by a ring being pulled off the finger with ease when the hand is elevated; also by the swellings of the feet that occur in relaxed habits, which swellings increase towards night, and subside in the morning, after the body has been in a horizontal posture for some hours.

In weak persons, the frequency of the pulse is increased by an erect posture, which may probably depend on gravity; as we know, from the observations of Macdonald and others, that an erect posture will make a difference of 15 or 20 beats in a minute. The experiments alluded to, were made by gently raising a person fastened to a board, where there being no muscular exertion, respiration would not be increased; so that the whole effect was probably owing to gravity accelerating the column of arterial blood.

The inverted posture produces a still more remarkable effect in accelerating the pulse, than the erect, for it sometimes causes it to beat 10 or 12 times more in the former case than in the latter.

While we are on this subject, it may not be improper to take notice of the effects of swinging on the circulation, which have been found by Dr. Carmichael Smyth, and others, to diminish the strength and velocity to such a degree, as to bring on fainting. These effects have never been satisfactorily accounted for; but they would seem to admit of an easy explanation on mechanical principles: they are undoubtedly owing, at least in a great measure, to the centrifugal force acquired by the blood.

By a centrifugal force, I mean, the tendency which revolving bodies have to fly off from the centre, which arises from their tendency to move in a straight line, agreeably to the laws of motion. Hence a tumbler of water may be whirled in a circle vertically without spilling it; the centrifugal force pushing the water against the bottom of the tumbler. In the same manner when the human body is made to revolve vertically in the arch of a circle, this centrifugal force will propel the blood from the head and heart towards he extremities; hence the circulation of the blood will be weakened, and the energy of the brain diminished. The contrary, however, will take place on a horizontal swing, as I have frequently observed, both on myself and others; for the centrifugal force in this case will propel the blood from the extremities towards the head.

It has been already observed, that the pulsations of the artery which we feel at the wrist, are occasioned by its alternate dilatations and contractions, which vary according to the strength and regularity of the circulation, which is liable to be affected by the smallest changes in the state of health. Hence physicians make use of the pulse as a criterion whereby to judge of the health of the body. And we may observe that there are few more certain characteristics of the state of the body than the pulse; yet the conclusions that have been drawn from it have often been erroneous; and this has arisen from trusting to observation without the aid of reason.

That we may better understand the phenomena of the pulse, I shall lay down the following postulata. 1st. It is now generally believed, that every part of the arterial system is endowed with irritability, or a power of contracting on the application of a stimulus, and that the blood acting on this contractibility, if the term may be allowed, causes contraction; and that the alternate relaxation and contraction gives the phenomenon pulsation. 2d. The greater the action of the stimulus of the blood, the greater will be the contraction, that is, the nearer will the sides of the artery approach towards the axis. 3d. That the velocity with which a muscular fibre, in a state of debility, contracts, is at least equal to that with which a fibre in a state of strength contracts, is a fact generally allowed by physiologists.

We shall afterwards see, that a deficient action of stimulus on the vessels may arise, either directly from diminishing the quantity of blood contained in them, or indirectly, from the application of too great a stimulant power, which has diminished the capability of contracting inherent in the vessels.

From these postulata, it will be evident, that the greater the action of the arteries, that is, the more powerful their contraction, the longer will be the intervals between the pulsations.

For the velocity being at least equal in debility and in strength, the times between the pulsations will be proportioned to the approach of the sides of the artery towards its axis: but the approach of the sides towards the axis is greater when the arteries are in a state of vigour than when debilitated; consequently the intervals between the pulsations will be greater when the arteries are in a state of vigour than when debilitated.

Hence it is evident, that a frequency of pulse must generally indicate a diminished action or debility; while a moderate slowness indicates a vigorous or just action.

Hence likewise the opinion of increased action, which has been supposed to take place in fevers, because a frequent pulse was observed, must be false, because the frequency arises from a directly opposite state, and indicates a diminished action of the vascular system.

In a sound and adult man the frequency of the pulse is about seventy beats in a minute; and in an infant, within the first five or six months, the pulse is seldom less than one hundred and twenty, and diminishes in frequency as the child grows older. But though seventy beats in the minute may be taken as a general standard; yet in persons of irritable constitutions the frequency is greater than this, and many, who are in the prime of life, have the pulse only between fifty and sixty.

It is generally observed, that the pulse is slower in the morning, that it increases in frequency till noon, after dinner it again becomes slow, and in the evening its frequency returns, which increases till midnight.

These phenomena may be rationally explained on the principles just laid down. When we rise in the morning, the contractibility being abundant, the stimulus of the blood produces a greater effect, the pulse becomes slow, and the contractions strong; it becomes more frequent, however, till dinner time, from a diminished contractibility; after dinner, from the addition of the stimulus of food and chyle, it again decreases in frequency, and becomes slow till the evening, when its frequency returns, because the contractibility becomes exhausted: and this frequency continues till the vital power have been recruited by sleep.

By the same principles it is easy to explain the quickness of the pulse in infancy, its gradual decrease till maturity, its slowness and strength during the meridian of life, and the return of its frequency during the decline.

Having now described the phenomena of the circulation, it will be proper to examine the changes produced by this function on the blood; and, in the first place, it may be observed, that the blood which returns by the vena cava to the heart, is of a dark colour inclining to purple; while that which passes from the left ventricle into the arteries, is of a bright vermilion hue. The blood which is found in the pulmonary artery has the same dark purple colour with that in the vena cava, while that in the pulmonary vein resembles the aortal blood in its brightness. Hence it would appear, that the blood, during its passage through the lungs, has its colour changed from a dark purple to a bright vermilion, in which state it is brought by the pulmonary vein to the left auricle of the heart; this auricle, contracting, expels the blood into the corresponding ventricle, by whose action, and that of the arteries, it is distributed to all parts of the body. When it returns, however, by the veins, it is found to have lost its fine bright colour. It would appear, therefore, that the blood obtains its red colour during its passage through the lungs, and becomes deprived of it during its circulation through the rest of the body.

That the blood contains iron, may be proved by various experiments: if a quantity of blood be exposed to a red heat in a crucible, the greatest part will be volatilised and burnt; but a quantity of brown ashes will be left behind, which will be attracted by the magnet. If diluted sulphuric acid be poured on these ashes, a considerable portion of them will dissolve; if into this solution we drop tincture of galls, a black precipitate will take place, or if we use prussiate of potash, a precipitate of prussian blue will be formed. These facts prove, beyond doubt, that a quantity of iron exists in the blood.

I shall not now particularly inquire how it comes there; it may partly be taken into the blood along with the vegetable and animal food, which is received into the stomach; for the greatest part of the animal and vegetable substances, which we receive as food, contain a greater or less quantity of iron. Or it may be partly formed by the animal powers, as would appear from the following circumstance. The analysis of an egg, before incubation, affords not the least vestige of iron, but as soon as the chick exists, though it has been perfectly shut up from all external communication, if the egg be burnt, the ashes will be attracted by the magnet.

But, however we may suppose the blood to obtain its iron, it certainly does contain it; if the coagulable lymph and serum of the blood be carefully freed from the red particles, by repeated washing, the strictest analysis will not discover in either of them a particle of iron, while the red globules thus separated will be found to contain a considerable quantity of this metal.

That the red colour of the blood depends upon iron, appears likewise from the experiments of Menghini, which show, that the blood of persons who have been taking chalybeate medicines for some time, is much more florid that it is naturally; the same is agreeable to my own observation. A late analysis by Fourcroy, has likewise proved, that the red colour of the blood resides in the iron; but, though the red colour of the blood may reside in the iron which it contains, we shall find that this colour is likewise connected with oxidation.

If the dark coloured blood, drawn from the veins, be put under a vessel containing oxygen gas, its surface will immediately become florid, while the bulk of the gas will be diminished. Mr. Hewson enclosed a portion of a vein between two ligatures, and injected into it a quantity of oxygen gas; the blood, which was before dark coloured, instantly assumed the hue of arterial blood. Thuvenal put a quantity of arterial blood under the receiver of an air pump; on exhausting the air it became of the dark colour of venous blood; on readmitting the air, it became again florid. He put it under a receiver filled with oxygen gas, and found the florid colour much increased.

Dr. Priestly exposed the blood of a sheep successively to oxygen gas, atmospheric air, and carbonic acid gas; and found, that in oxygen gas its colour became very florid, less so in atmospheric air, and in carbonic acid gas it became quite black. He filled a bladder with venous blood, and exposed it to oxygen gas; the surface in contact with the bladder immediately became florid, while the interior parts remained dark coloured.

All these facts prove, that the red colour which the blood acquires in the lungs, is owing to the oxygen, which probably combines with it, and the last mentioned fact shows, that oxygen will act on the blood, even though a membrane similar to the bladder, be interposed between them.

The same effect, probably, takes place in the lungs; the blood is circulated through that organ by a number of fine capillary arteries; and it is probable that the oxygen acts upon the blood through the membranes of these arteries, in the same manner that it does through the bladder.

In short, it seems likely, that the blood, during its circulation through the lungs, becomes combined with oxygen; that this oxidated blood, on its return to the heart, is circulated by the arteries to all parts of the body; and that, during this circulation, its oxygen combines with the hydrogen and carbon of the blood, and perhaps with those parts of the body with which it comes into contact; it is therefore brought back to the heart, by the veins, of a dark colour, and deprived of the greatest part of its oxygen.

This is the most probable theory, in the present state of our knowledge; it was proposed by Lavoisier, who imagines the focus of heat, or fireplace to warm the body, to be in the lungs: others, however, have thought it more consonant to facts, to suppose, that, instead of the oxygen uniting with carbon and hydrogen in the lungs, and there giving out its heat, the oxygen is absorbed by the blood, and unites with these substances during the circulation, so that heat is produced in every part of the body; and this doctrine seems certainly supported by several facts and experiments.

The circulation of the blood, though so simple and beautiful a function, was unknown to the ancient physicians, and was first demonstrated by our countryman, Harvey; when he first published his account of this discovery, he met with the treatment which is generally experienced by those who enlighten and improve the comfort of their fellow creatures, by valuable discoveries. The novelty and merit of this discovery drew upon him the envy of most of his contemporaries in Europe, who accordingly opposed him with all their power; and some universities even went so far, as to refuse the honours of medicine to those students, who had the audacity to defend this doctrine; but afterwards, when they could not argue against truth and conviction, they attempted to rob him of the discovery, and asserted that many of the ancient physicians, and particularly Hippocrates, were acquainted with it. Posterity, however, who can alone review subjects of controversy without prejudice, have done ample justice to his memory.

LECTURE IV.
DIGESTION, NUTRITION, &c.

The human body, by the various actions to which it is subject, and the various functions which it performs, becomes, in a short time, exhausted; the fluids become dissipated, the solids wasted, while both are continually tending towards putrefaction Notwithstanding which, the body still continues to perform its proper functions, often for a considerable length of time; some contrivance, therefore, was necessary to guard against these accelerators of its destruction. There are two ways in which the living body may be preserved; the one by assimilating nutritious substances, to repair the loss of different parts; the other to collect, in secretory organs, the humours secreted from these substances.

We are admonished of the necessity of receiving substances into the body, to repair the continual waste, by the appetites of hunger and thirst. For the stomach being gradually emptied of its contents, and the body, in some degree, exhausted by exercise, we experience a disagreeable sensation in the region of the stomach, accompanied by a desire to eat, at first slight, but gradually increasing, and at last growing intolerable, unless it be satisfied.

When the fluid parts have been much dissipated, or when we have taken, by the mouth, any dry food, or acrid substance, we experience a sensation of heat in the fauces, and at the same time a great desire of swallowing liquids. The former sensation is called hunger, and the latter thirst.

From the back part of the mouth passes a tube, called the oesophagus or gullet, its upper end is wide and open, spread behind the tongue to receive the masticated aliment: the lower part of this pipe, after it has passed through the thorax, and pierced the diaphragm, enters the stomach, which is a membranous bag, situated under the left side of the diaphragm: its figure nearly resembles the pouch of a bagpipe, the left end being most capacious; the upper side is concave, and the lower convex: it has two orifices, both on its upper part; the left, which is a continuation of the oesophagus, and through which the food passes into the stomach, is named cardia; and the right, through which the food is conveyed out of the stomach, is called pylorus: within this last orifice is a circular valve, which, in some degree, prevents the return of the aliment into the stomach.

From the pylorus, or right orifice of the stomach, arise the intestines, or bowels, which consist of a long and large tube, making several circumvolutions, in the cavity of the abdomen; this tube is about five or six times as long as the body to which it belongs. Though it is one continued pipe, it has been divided, by anatomists, into six parts, three small, three large. The three small intestines are the duodenum, the jejunum, and the ileum; the duodenum commences at the pylorus, and is continued into the jejunum, which is so called from its being generally found empty: the ileum is only a prolongation of the jejunum, and terminates in the first of the great intestines, called the caecum. The other great guts are the colon and the rectum.

The whole of what has been described is only a production of the same tube, beginning at the oesophagus. It is called by anatomists the intestinal canal, or prima via, because it is the first passage of the food. It has circular muscular fibres, which give it a power of contracting when irritated by distension; and this urges forward the food which is contained in it. This occasions a worm like motion of the whole intestines, which is called their peristaltic motion.

The mesentery is a membrane beginning loosely on the loins, and thence extending to all the intestines; which it preserves from twisting by their peristaltic motion. It serves also to sustain all the vessels going to and from the intestines, namely the arteries, veins, lacteals, and nerves; it also contains several glands, called, from their situation, mesenteric glands.

The lacteal vessels consist of a vast number of fine pellucid tubes, which arise by open mouths from the intestines, and proceeding thence through the mesentery, they frequently unite, and form fewer and larger vessels, which pass through the mesenteric glands, into a common receptacle or bag, called the receptacle of the chyle.

The use of these vessels is to absorb the fluid part of the digested aliment, called chyle, and convey it into the receptacle of the chyle, that it may be thence carried through the thoracic duct into the blood.

The receptacle of the chyle is a membranous bag, about two thirds of an inch long, and one third of an inch wide, at its superior part it is contracted into a slender membranous pipe, called the thoracic duct, because its course is principally through the thorax; it passes between the aorta and the vena azygos, then obliquely over the oesophagus, and great curvature of the aorta, and continuing its course towards the internal jugular vein, it enters the left subclavian vein on its superior part.

There are several other viscera besides those I have described, which are subservient to digestion; among these may be mentioned the liver, gall bladder, and pancreas. The liver is the largest gland in the body, and is situated immediately under the diaphragm, principally on the right side. Its blood vessels that compose it as a gland, are the branches of the vena portarum, which, as I mentioned in the last lecture, enters the liver and distributes its blood like an artery. From this blood the liver secretes the bile, which is conveyed by the hepatic duct, towards the intestines: before this duct reaches the intestines, it is joined by another, coming from the gall bladder: these two ducts uniting, form a common duct, which enters the duodenum obliquely, about four inches below the pylorus of the stomach.

The gall bladder, which is a receptacle of bile, is situated between the stomach and the liver; and the bile which comes from the liver, along the hepatic duct, partly passes into the duodenum, and partly along the cystic duct into the gall bladder. When the stomach is full, it presses on the gall bladder, which will squeeze out the bile into the duodenum at the time when it is most wanted.

The bile is a thick bitter fluid, of a yellowish green colour, composed chiefly of soda and animal oil, forming a soap; and it is most probably in consequence of this saponaceous property that it assists digestion, by causing the different parts of the food to unite together by intermediate affinity. When the bile is prevented from flowing into the intestines, by any obstruction in the ducts, digestion is badly performed, costiveness takes place, and the excrements are of a white colour, from being deprived of the bile. This fluid, stagnating in the gall bladder, is absorbed by the lymphatics, and carried into the blood, communicating to the whole surface of the body a yellow tinge, and other symptoms of jaundice.

The jaundice therefore is occasioned by an obstruction to the passage of the bile into the intestines, and its subsequent absorption into the blood: this obstruction may be caused either by concretions of the bile, called gall stones, or by a greater viscidity of the fluid, or by a spasm, or paralysis of the biliary ducts.

The pancreas, or sweet bread, is a large gland lying across the upper and back part of the abdomen, near the duodenum. It has a short excretory duct, about half as wide as a crow quill, which enters the duodenum at the same place where the bile duct enters it.

The food being received into the mouth, is there masticated or broken down, by the teeth, and impregnated with saliva, which is pressed out of the salivary glands, by the motions of the jaw and the muscles of the mouth. It then descends, through the oesophagus, into the stomach, where it becomes digested, and, in a great measure, dissolved, by the gastric juice, which is secreted by the arteries of the stomach. It is then pushed through the pylorus, or right orifice of the stomach into the duodenum, where it becomes mixed with the bile from the gall bladder and liver, and the pancreatic juice from the pancreas. These fluids seem to complete the digestion: after this, the food is continually moved forwards by the peristaltic motion of the intestines.

The chyle, or thin and milky part of the aliment, being absorbed by the lacteals, which rise, by open mouths, from the intestines, is carried into the receptacle of the chyle, and from thence the thoracic duct carries it to the subclavian vein, where it mixes with the blood, and passes with it to the heart.

The food of animals is derived from the animal or vegetable kingdoms. There are some animals which eat only vegetables, while others live only on animal substances. The number and form of the teeth, and the structure of the stomach, and bowels, determine whether an animal be herbivorous, or carnivorous. The first class have a considerable number of grinders, or dentes molares; and the intestines are much more long and bulky; in the second class, the cutting teeth are predominant, and the intestines are much shorter.

Man seems to form an intermediate link between these two classes: his teeth, and the structure of the intestines, show, that he may subsist both on vegetable and animal food; and, in fact, he is best nourished by a proper mixture of both. This appears from those people who live solely on vegetables, as the Centoo tribes, and those who subsist solely on animals, as the fish eaters of the northern latitudes, being a feebler generation than those of this country, who exist on a proper mixture of both. A due proportion, therefore, of the two kinds of nourishment, seems undoubtedly the best.

Having taken a general view of the course of the aliment into the blood, I shall now examine more particularly, how each part of the organs concerned in digestion, or connected with that function, contributes to that end.

The food being received into the mouth, undergoes various preparations, which fit it for those changes it is afterwards to undergo. By the teeth the parts of it are divided and ground, softened and liquified by the saliva, and properly compressed by the action of the tongue and mastication.

The mouth, in most animals, is armed with very hard substances, which, by the motion of the lower jaw, are brought strongly into contact. Those parts of the teeth which are above the sockets, are not simply bony, they are much harder than the bones, and possess the property of resisting putrefaction, as long as this hard crust continues to cover them. The teeth are divided into three classes: 1st. The cutting teeth, which are sharp and thin, and which serve to cut or divide the food: 2nd. The canine teeth, which serve to tear it into pieces still smaller: 3rd. The grinders, which present large and uneven surfaces, and actually grind the substance already broken down by the other teeth. Birds, whom nature has deprived of teeth, have a strong muscular stomach, called the gizzard, which serves the purposes of teeth, and they even take into the stomach small pieces of grit, to assist in grinding to a powder the grain that they have swallowed.

Among those parts of the mouth which contribute to the preparation of the food, we must reckon the numerous glands which secrete saliva, and which have therefore been called salivary glands. The saliva is a saponaceous liquor, destitute of taste or smell, which is squeezed out from these glands, and mixed with the food during mastication. In the mouth, therefore, the food becomes first broken down by the teeth, impregnated with saliva, and reduced to a soft pasty substance, capable of passing with these, through the oesophagus, into the stomach. It is here that it undergoes the change, which is particularly termed digestion.

Digestion comprehends two classes of phenomena, distinct from each other: 1st. Physical and chemical: 2nd. Organic and vital. The object of the first, is to bring the alimentary substances to such a state as is necessary, that they may be capable of the new combinations into which they are to enter, to obtain the animal character. The object of the second is, to produce those combinations which some have thought to be very different from those produced by simple chemical attractions.

The physical and chemical phenomena of digestion, relate chiefly, 1st. To the action of heat; 2ndly. To the dissolution of the alimentary substances. The heat of the animal is such, as is well fitted to promote solution.

That digestion is performed by solution, is evident, from several experiments, particularly those made by Dr. Stevens, who enclosed different alimentary substances in hollow spheres of silver, pierced with small holes. These were swallowed, and after remaining some time in the stomach, the contents were found dissolved. The great agent of solution is the gastric juice, which possesses a very strong solvent power. This juice is secreted by the arteries of the stomach; it may be collected in considerable quantity, by causing an animal that has been fasting for some time, to swallow small hollow spheres, or tubes of metal filled with sponge.

This liquid does not act indiscriminately upon all substances; for if grains of corn be put into a perforated tube, and a granivorous bird be made to swallow it, the corn will remain the usual time in the stomach without alteration; whereas if the husk of the grain be previously taken off, the whole of it will be dissolved. There are many substances likewise which pass unaltered through the intestines of animals, and consequently are not acted upon by the gastric juice. This is the case frequently with grains of oats, when they have been swallowed by horses entire, with their husks on. This is the case likewise with the seeds of apples and other fruits, when swallowed entire by man; yet if these substances have been previously ground by the teeth, they will be digested. It would appear therefore, that it is chiefly the husk or outside of these substances which resists the action of the gastric juice.

This juice is not the same in all animals; for many animals cannot digest the food on which others live. Thus sheep live wholly on vegetables, and if they are made to feed on animals, their stomachs will not digest them: others again, as the eagle, feed wholly on animal substances, and cannot digest vegetables.

The accounts of the experiments made on gastric juice are very various: sometimes it has been found of an acid nature, at other times not. The experiments of Spallanzani show, however, that this acidity is not owing to the gastric juice, but to the food. The result of his experiments, which have been very numerous, prove, that the gastric juice is naturally neither acid nor alkaline. No conclusion, however, can be drawn from these experiments made out of the stomach, with respect to the nature of the gastric juice; nor do the analyses which have been made of it throw any light on its mode of action. But, from the experiments which have been made on digestion, in the stomach, particularly by Spallanzani, the following facts have been established.

The gastric juice attacks the surfaces of bodies, and combines chemically with their particles. It operates with more energy and rapidity, the more the food is divided, and its action increased by a warm temperature. By the action of digestion, the food is not merely reduced to very minute parts, but its chemical properties become changed; its sensible properties are destroyed, and it acquires new and very different ones. This juice does not act as a ferment; so far from it, it is a powerful antiseptic, and even restores flesh which is already putrid.

When the alimentary substances have continued a sufficient time in the stomach, they are pushed into the intestines, where they become mixed with the bile and pancreatic juice, as was before observed. What changes are caused by these substances, we have yet to learn; but there is no doubt, that they serve some important purposes. By the peristaltic motion of the bowels, the alimentary matters thus changed are carried along, and applied to the mouths of the lacteal vessels, which open into the intestines, like a sponge, and by some power, not well understood, absorb that part which is fitted for assimilation, while the remainder is rejected as an excrement.

The lacteal vessels are furnished with valves, which allow a free passage to the chyle from the intestines, but prevent its return. The most inexplicable thing in this operation, is the power which these vessels possess of selecting from the intestinal mass, those substances which are proper for nutrition, and rejecting those which are not.

These lacteal vessels, as was before observed, pass through the mesentery, and their contents seem to undergo some important change in the mesenteric glands. The chyle which passes through vessels, appears to be an oily liquor, less animalised than milk, and its particles seem to be held in solution by the intermedium of a mucilaginous principle. It is conveyed along the thoracic duct in the manner already described, and enters the blood slowly, and, as it were, drop by drop, by the subclavian vein; in this way it becomes intimately mixed with the blood, and combining with oxygen in the lungs, it acquires a fibrous character, and becomes fit to nourish the body.

We have now seen how the process of digestion is performed, at least, so far as we are acquainted with it, and how its products are conveyed into the blood. But to what purposes the blood is employed, which is formed with so much care, we have yet to discover. It seems to answer two purposes. The parts of which the body is composed, namely, bones, muscles, ligaments, membranes, &c. are continually changing: in youth they are increasing in size and strength, and in mature age they are continually acting, and, consequently, continually liable to waste and decay. They are often exposed to accidents, which render them unfit for performing their various functions; and even when no such accidents happen, it seems necessary for the health of the system that they should be perpetually renewed. Materials must therefore be provided for repairing, increasing, or renewing all the various organs of the body. The bones require phosphate of lime, and gelatine, the muscles fibrine, and the cartilages and membranes albumen; and accordingly we find all these substances contained in the blood, from whence they are drawn, as from a storehouse, whenever they are wanted. The process by which these different parts of the blood become various parts of the body is called assimilation.

Over the nature of assimilation the thickest darkness still hangs; all that we know for certain is, that there are some conditions necessary to its action, without which it cannot take place. These are, 1. A sound and uninterrupted state of the nerves. 2. A sound state of the blood vessels. 3. A certain degree of tone or vigour in the vessels of the part.

There remains yet to be noticed another set of vessels, connected with the circulating and nutritive systems, called lymphatics. These vessels are very minute, and filled with a transparent fluid: they rise by open mouths in every cavity of the body, as well as from every part of the surface, and the course of those from the lower extremities, and indeed from most of the lower parts of the body, is towards the thoracic duct, which they enter at the same time with the lacteal vessels already described. They are furnished, like the lacteals, with numerous valves, which prevent their contents from returning towards their extremities.

The minute arteries in every part of the body exhale a colourless fluid, for lubricating the different parts, and other important purposes: and the lymphatic vessels absorb the superfluous quantity of this fluid, and convey it back to the blood.

It must be evident therefore, that, if the lymphatics in any cavity become debilitated, or by any other means be prevented from absorbing this exhaled fluid, an accumulation of it will take place: the same will happen, if the exhaling arteries be debilitated, so as to allow a greater quantity of fluid to escape than the absorbents can take up. When the balance between exhalation and absorption is destroyed, by either or both of these means, a dropsy will be the consequence.

Before we finish the subject of digestion, I shall take a short view of some of the morbid affections, attending this important function of the animal economy.

A deficiency of appetite may arise, either from an affection of the stomach, or a morbid state of the body: for there is such a sympathy between the stomach and the rest of the system, that the first is very seldom disordered, without communicating more or less disorder to the system: nor can the system become deranged and the stomach remain sound.

A want of appetite may arise from overloading the stomach, whereby its digestive powers will be weakened. And this may be occasioned in two ways. First, by taking food of the common quality in too great quantity, which will certainly weaken the powers of the stomach. An excellent rule, and one which if more attended to, would prevent the dreadful consequences of indigestion, is always to rise from the table with some remains of appetite. This is a rule applicable to every constitution, but particularly to the sedentary and debilitated.

The second way in which the stomach may be debilitated, is by taking food too highly stimulating or seasoned; and this even produces much worse effects than an over dose with respect to quantity. The tone of the stomach is destroyed, and a crude unassimilated chyle is absorbed by the lacteals, and carried into the blood, contaminating its whole mass. Made dishes, enriched with hot sauces, stimulate infinitely more than plain food, and bring on diseases of the worst kind: such as gout, apoplexy, and paralysis. "For my part," says an elegant writer, "when I behold a fashionable table set out in all its magnificence, I fancy I see gouts, and dropsies, fevers, and lethargies, with other innumerable distempers, lying in ambuscade among the dishes."

All times of the day are not equally fitted for the reception of nourishment. That digestion may be well performed, the functions of the stomach and of the body must be in full vigour. The early part of the day therefore is the proper time for taking nutriment; and, in my opinion, the principal meal should not be taken after two or three o'clock, and there should always be a sufficient time between each meal to enable the stomach to digest its contents. I need not remark how very different this is from the common practice of jumbling two or three meals together, and at a time of the day likewise when the system is overloaded. The breakfast at sunrise, the noontide repast and the twilight pillow, which distinguished the days of Elizabeth, are now changed for the evening breakfast, and the midnight dinner. The evening is by no means the proper time to take much nourishment: for the powers of the system, and particularly of the stomach, are then almost exhausted, and the food will be but half digested. Besides, the addition of fresh chyle to the blood, together with the stimulus of food acting on the stomach, always prevents sleep, or renders it confused and disturbed, and instead of having our worn out spirits recruited, by what is emphatically called by Shakespeare, "the chief nourisher in life's feast," and rising in the morning fresh and vigorous, we become heavy and stupid, and feel the whole system relaxed.

It is by no means uncommon, for a physician to have patients, chiefly among people of fashion and fortune, who complain of being hot and restless all night, and having a bad taste in the mouth in the morning. On examination, I have found that, at least in nineteen cases out of twenty, this has arisen from their having overloaded their stomachs, and particularly from eating hot suppers; nor do I recollect a single instance of a complaint of this kind in any person not in the habit of eating such suppers.

The immoderate use of spirituous and fermented liquors, is still more destructive of the digestive powers of the stomach; but this will be better understood, when we have examined the laws by which external powers act upon the body. The remarks I have made could not, however, I think, have come in better, than immediately after our examination of the structure of the digestive organs, as the impropriety of intemperance, with respect to food, is thus rendered more evident.

The appetite becomes deficient from want of exercise, independently of the other causes that have been mentioned. Of all the various modes of preserving health, and preventing diseases, there is none more efficacious than exercise; it quickens the motion of the fluids, strengthens the solids, causes a more perfect sanguification in the lungs, and, in short, gives strength and vigour to every function of the body. Hence it is, that the Author of nature has made exercise absolutely necessary to the greater part of mankind for obtaining means of existence. Had not exercise been absolutely necessary for our well being, says the elegant Addison, nature would not have made the body so proper for it, by giving such an activity to the limbs, and such a pliancy to every part, as necessarily produce those compressions, extensions, contortions, dilatations, and all other kinds of motion, as are necessary for the preservation of such a system of tubes and glands.

We may, indeed, observe, that nature has never given limbs to any animal, without intending that they should be used. To fish she has given fins, and to the fowls of the air wings, which are incessantly used in swimming and flying; and if she had destined mankind to be eternally dragged about by horses, her provident economy would surely have denied them legs.

The appetite becomes deficient on the commencement of many diseases, but this is to be looked upon here rather as a salutary than as a morbid symptom, and as a proof that nature refuses the load, which she can neither digest nor bear with impunity.

In healthy people the appetite is various, some requiring more food than others; but it sometimes becomes praeternaturally great, and then may be regarded as a morbid symptom. The appetite may be praeternaturally increased, either by an unusual secretion of the gastric juice, which acts upon the coats of the stomach, or by any acrimony, either generated in, or received into the stomach, or, lastly, by habit, for people undoubtedly may gradually accustom themselves to take more food than is necessary.

The appetite sometimes becomes depraved, and a person thus affected, feels a desire to eat substances that are by no means nutritious, or even esculent: this often depends on a debilitated state of the whole system. There are some instances, however, in which this depravity of the appetite is salutary; for example, the great desire which some persons, whose stomachs abound with acid, have for eating chalk, and other absorbent earths: likewise, the desire which scorbutic patients have for grass, and other fresh vegetables. Appetites of this kind, if moderately indulged in, are salutary, rather than hurtful.

The appetite for liquids as well as solids is sometimes observed to be deficient, and sometimes too great. The former can scarcely be considered as a morbid symptom, provided the digestion and health be otherwise good. But when along with diminished thirst, the fauces and tongue are dry, this deficiency may be regarded as a morbid and dangerous symptom.

A more common morbid symptom, however, is too great thirst, which may arise from a deficiency of fluids in the body, produced by violent exercise, perspiration, too great a flow of urine, or too great an evacuation of the intestines. A praeternatural thirst may likewise arise from any acrid substance received into the stomach, which our provident mother, nature, teaches us to correct by dilution; this is the case with respect to salted meats, or those highly seasoned with pepper. It may arise also from the stomach being overloaded with unconcocted aliment, or from a suppressed or diminished secretion of the salivary liquors in the mouth, which may arise from fever, spasm, or affections of the mind; an increased thirst may likewise take place, from a derivation or determination of the fluids to other parts of the body; of this, dropsy affords an example. Indeed, various causes may concur to increase the thirst; this is the case in most fevers, where great thirst is occasioned by the dissipation of the fluids of the body by heat, as well as by the diminished secretion of the salivary humours which should moisten the mouth; to which may be added, the heat and diminished concoctive powers of the stomach.

From what has been said, we can easily understand, why praeternatural thirst may sometimes be a necessary instinct of nature, at other times, an unnecessary craving; why acids, acescent fruits, and weak fermented liquors quench thirst more powerfully than pure water; and lastly, why thirst, in some instances, may be relieved by emetics, when it has resisted other remedies.

There is no organ of the body whose functions are so easily deranged as those of the stomach; and these derangements prove a very fertile source of disease; they ought, therefore, carefully to be guarded against; and it is fortunate for us that we have this generally in our power, if we would but avail ourselves of it: for most of the derangements proceed from the improper use of food and drink, and a neglect of exercise. Indeed, when we examine, we shall find but a short list in the long catalogue of human diseases, which it is not in our power to guard against and prevent: and which surely will be guarded against, when their causes are known, and consequences understood.

Among the diseases arising from a disordered state of the stomach and indigestion, may be enumerated the following: great oppression and anxiety, pain in the region of the stomach, with acid eructations, nausea, vomiting, the bowels sometimes costive, sometimes too loose, but seldom regular, depression of spirits, and all the long list, commonly, but very improperly, termed nervous complaints, deficient nutrition, and consequently general weakness, a relaxed state of the solids, too great a tenuity of the fluids, headach, vertigo, and many other complaints, too numerous to mention here.

The greatest misfortune, and which indeed arises from a want of physiological knowledge, is, that people labouring under these disorders, imagine they may be cured by the reception of drugs into the stomach, and thus they are induced to receive into that organ, half the contents of an apothecary's shop. There is no doubt that these complaints may oftentimes be alleviated, and the cure assisted, by medicines: thus, when the stomach is overloaded, this may be removed by an emetic; the same complaint of the bowels may be removed by a cathartic; and when the stomach is debilitated, we are acquainted with some substances which will give it vigour, such as iron, the Peruvian bark, and several kinds of bitters. These however, when used alone, afford but temporary relief; and unless the cause which induced the disease be removed, it will afterwards return with redoubled violence. When the stomach, for instance, is debilitated by want of exercise, I would ask, is there an article in the whole materia medica, that can cure the complaints of sedentary people, unless proper exercise at the same time be taken? With exercise tonic remedies will undoubtedly accelerate the cure, but without it, they will only make bad worse.

Again, when the stomach is debilitated by the use of improper food, or the abuse of fermented or spirituous liquors, I would say to any one who pretended to cure me of these complaints, without my making a total change in the manner of living, that he either was ignorant of the matter, or intended to deceive me.

In many cases the change of food must be strictly observed and persevered in for a long time before a cure can be effected. In some instances where the powers of the stomach were too weak to prevent the food from undergoing perhaps both a vinous and acetous fermentation, and where, in consequence of the disengagement of gas and the formation of acid, the most excruciating pains were felt, the most dreadful sickness experienced, and all the symptoms of indigestion present in the most aggravated state; after almost every article in the materia medica, generally employed, had been tried without success, I have cured the patient merely by prohibiting food subject to fermentation, such as vegetables, and enjoining a strict use of animal food alone.

In short, wherever the cause of a disease can be ascertained, the grand and simple secret in the cure, is the careful removal of that cause.

LECTURE V. OF THE SENSES IN GENERAL.

In this lecture, I propose to take a view of the connexion of man with the external world, and shall endeavour to point out the manner in which he becomes acquainted with external objects, by means of the faculties called senses.

A human creature is an animal endowed with understanding, and reason; a being composed of an organized body, and a rational mind.

With respect to his body, he is pretty similar to other animals, having similar organs, powers, and wants. All animals have a body composed of several parts, and, though these may differ from the structure of the human body in some circumstances, to accommodate it to peculiar habits and wants of the animal, still there is a great similarity in the general structure.

The human body is feeble at its commencement, increases gradually in its progress by the help of nourishment and exercise, till it arrives at a certain period, when it appears in full vigour; from this time it insensibly declines to old age, which conducts it at length to dissolution. This is the ordinary course of human life, unless it happens to be abridged either by disease or accident.

With regard to his reasoning faculties, or mind, man is eminently distinguished from other animals. It is by this noble part that he thinks, and is capable of forming just ideas of the different objects that surround him: of comparing them together; of inferring from known principles unknown truths; of passing a solid judgment on the mutual agreement of things, as well as on the relations they bear to him; of deliberating on what is proper or improper to be done; and of determining how to act. The mind recollects what is past, joins it with the present, and extends its views to futurity. It is capable of penetrating into the causes of events, and discovering the connexion that exists between them.

Governed by invariable laws, which connect him with all the beings, whether animate or inanimate, among which he exists, man has certain relations of convenience, and inconvenience, arising from the particular constitution of the surrounding objects, as well as of his own body. These external objects possess qualities which may be useful or prejudicial to him; and his interest requires, that he should be capable of ascertaining and appreciating these properties.

It is by sensation, or feeling, that the knowledge of external objects is obtained. The faculty of feeling, modified in every organ, perceives those qualities for which the peculiar structure of the organ is fitted; and all the various sensations of sound, colour, taste, smell, resistance, and temperature, find appropriate organs by which they are perceived, without mixing with, or confounding each other. External objects, therefore, act upon the parts of the body endowed with feeling, and their action is diversified in such a manner, as to give us a great number of sensations, which appear to have no resemblance to each other, and which make us acquainted with the various properties of surrounding objects.

It would not, however, have been sufficient for man, merely to have possessed this power of perceiving the different properties of the objects which surround him: it was necessary likewise, that he should be possessed of motion, that he might be able to approach or avoid them, to seize or repulse them, as it suited his convenience or advantage. By the extreme mobility of his limbs, he is able to move his body, and transport it from place to place; to bring external objects nearer to him, to remove them to a greater distance, and to place them in such situations and such circumstances, as may enable them to act on each other, and produce the changes which he wishes.

The human body, therefore, may be regarded as a machine composed (besides the moving parts which have formerly been noticed) of divers organs upon which external objects act, and produce those impressions which convince us of their presence, and make us acquainted with their properties. These impressions are transmitted to the sentient principle, or mind; and the faculty we possess of perceiving these impressions has been called by physiologists, sensibility.

Sensation has generally been defined by metaphysicians to be a change in the mind, of which we are conscious, caused by a correspondent change in the state of the body. This definition, however, leaves the matter where they found it, and throws no light whatever on the nature of sensation; nor can we say any thing more concerning it, than that, when the organs are in a sound state, certain sensations are perceived, which force us to believe in the existence of external objects, though there is no similarity whatever, nor any necessary connexion, that we can perceive, between the sensation and the object which caused it.

All the different degrees of sensation may be reduced to two kinds: pleasant and painful. The nature of these two primitive modes of sensation, is as little known to us as their different species: all that can be said, is, that the general laws by which the body is governed, are such, that pleasure is generally connected with those impressions which tend to its preservation, and pain with those which cause its destruction.

In a general point of view, sensibility may be regarded as an essential property of every part of the living body, disposing each part to perform those functions, the object of which is to preserve the life of the animal. Sensibility presides over the most necessary functions, and watches carefully over the health of the body: she directs the choice of the air proper for respiration, and also of alimentary substances; the mechanism of the secretions is likewise placed under her power; and in the same way that the eye perceives colours, and the ear sounds, so every animated and living part is fitted to receive impressions from the objects appropriated to it.

That every part of the animal is endowed with sensibility, is evident from a variety of facts, particularly from the action which follows when a muscle taken out of the animal body is irritated by any stimulus: this is evident, by a variety of facts mentioned by Whytt, Boerhaave, and others, which show, that parts recently taken from the animal body retain a portion of sensibility, which continues to animate them, and render them capable of action for a considerable time.

The primary organ of sensation appears to be the brain, its continuation in the form of medulla oblongata and spinal marrow, and the various nerves proceeding from these; and it seems now generally agreed, that unless there be a free communication of nerves between the part where the impression is made, and the brain, no sensation will take place; for instance, if the nerves be cut or compressed.

In a sound body, sensation is caused, whenever a change takes place in the state of the nervous power, whether that change be produced by an external, or an internal cause. The former kind of sensation is said to arise from impression or impulse, the latter from consciousness.

Every impression or impulse is not, however, equally calculated to produce sensation; for this purpose, a middle degree of impulse appears the best. An impulse considerably less produces no sensation, and one more violent may cause pain, but no proper sensation denoting the presence or properties of external objects. Thus too small a degree of light makes no impression on the optic nerve; and if the object be too strongly illuminated, the eye is pained, but has no proper idea of the figure or colour of the object. In the same way, if the vibrations which give us an idea of sound, be either too quick or too slow, we shall not obtain this idea. When the vibration is too quick, a very disagreeable and irritating sensation is perceived, as for instance, in the whetting of a saw: and on the other hand, when the vibrations are too slow, they will not produce a tone or sound. This might be proved of all the senses, and shows, that a certain degree of impression is necessary to produce perfect sensation.

There is another circumstance likewise requisite to produce sensation: it is not enough, that the impression should be of the proper strength; it is necessary likewise, that it should remain for some time, otherwise no sensation will be produced. There are many bodies whose magnitude is amply sufficient to be perceived by the eye; yet, by reason of their great velocity, the impulse they make on any part of the retina is so short, that they are not visible. This is proved by our not perceiving the motions of cannon and musket balls, and many other kinds of motion. On this principle depends the art of conjuration, or legerdemain; the fundamental maxim of those who practise them, is, that the motion is too quick for sight.

If the impulse be of a proper degree, and be continued for a sufficient length of time, the impression made by it will not immediately vanish with the impulse which caused it, but will remain for a time proportioned to the strength of the impulse. This, with respect to sight, is proved by whirling a firebrand in a circular manner, by which the impression of a circle is caused, instead of a moving point: and, with respect to hearing, it may be observed, that when children run with a stick quickly along railing, or when a drum is beaten quickly, the idea of a continued sound is produced, because the impression remains some time: for it is evident, that the sounds produced in succession are perfectly distinct and insulated.

Sensation likewise depends, in a great measure, on the state of the mind, and on the degree of attention which it gives. For if we are engaged in attention to any object, we are insensible of the impressions made upon us by others, though they are sufficiently strong to affect us at other times. Thus, when our attention is fixed strongly upon any particular object, we become insensible of the various noises that surround us, though these may be sometimes very loud. On the contrary, if our attention be upon the watch, we can perceive slight, and almost neglected impressions, while those of greater magnitude become insensible. The ticking of a clock becomes insensible to us from repetition, but if we attend to it, we become easily sensible of it, though at the same time we become insensible of much stronger impressions, such as the rattling of coaches in the streets.

The attention depends in some degree on the will, but is generally given to those impressions which are particularly strong, new, pleasant, or disagreeable; in short, to those which particularly affect the mind. Hence it is, that things which are new, produce the most vivid impressions, which gradually grow fainter, and at last become imperceptible.

There is one circumstance respecting sensation, which will probably account for our only perceiving those impressions to which the mind attends: and this is, that the mind is incapable of perceiving more than one impression at a time: the more accurately we examine this, the greater reason we shall have to think it true; but the mind can turn its attention so quickly, from one object to another, that at first sight, we are led to believe, that we are able to attend to several at the same time.

But though the mind cannot perceive or attend to various sensations at the same time, yet if two or more of these are capable of uniting in such a manner as to produce a compound sensation, this may be perceived by the mind.

This compound sensation may be produced either by impressions made at the same instant, or succeeding each other so quickly, that the second takes place before the first has vanished.

As an instance of the first, we may mention musical chords, or the sounds produced by the union of two or more tones at the same time. We have another instance likewise in odours or smells; if two or more perfumes be mixed together, a compound odour will be perceived, different from any of them.

As an instance of the latter, if a paper painted of various colours be made to revolve rapidly in a circle, a compound colour, different from any of them, will be perceived. These observations apply particularly to the senses we have mentioned, and likewise to taste: but the sensations afforded us by touch do not seem capable of being compounded in this manner.

There are many things necessary to perfect sensation, besides those that have been mentioned. The degree and perfection of sensation will depend much on the mind, and will be continually altered by delirium, torpor, sleep, and other circumstances; much likewise depends on the state of the organs with respect to preceding impressions; for if any organ of sense have been subjected to a strong impression, it will become nearly insensible of those which are weaker.

Of this innumerable instances may be given: an eye which has been subjected to a strong light, becomes insensible of a weaker: and on the contrary, if the organs of sense have been deprived of their accustomed impressions for some time, they are affected by very slight ones. Hence it is, that when a person goes from daylight into a darkened room, he can at first see nothing; by degrees however he begins to have an imperfect perception of the different objects, and if he remain long enough, he will see them with tolerable distinctness, though the quantity of light be the same as when he entered the room, when they were invisible to him.

Sensation often arises from internal causes, without any external impulse. To this source may be referred consciousness, memory, imagination, volition, and other affections of the mind. These are called the internal senses. The senses, whether internal or external, have never been accurately reduced to classes, orders, or genera; the external indeed are generally referred to five orders; namely, seeing, hearing, smelling, tasting, and feeling, or touch. With respect to the four first, the few qualities of external bodies which each perceives may be easily reduced to classes, each of which may be referred to its peculiar organ of sensation, because each organ is so constituted, that it can only be affected by one class of properties; thus the eye can only be affected by light; the ear by the vibrations of the air, and so of the rest.

The same organ, whatever be its state, or whatever be the degree of impulse, always gives to the mind a similar sensation; nor is it possible, by any means we are acquainted with, to communicate the sensation peculiar to one organ by means of another. Thus we are incapable, for instance, of hearing with our eyes, and seeing with our ears: nor have we any reason to believe that similar impressions produce dissimilar sensations in different people. The pleasure, however, as well as the pain and disgust, accompanying different sensations, differ very greatly in different persons, and even in the same person at different times; for the sensations which sometimes afford us pleasure, at other times produce disgust.

Habit has a powerful influence in modifying the pleasures of sensation, without producing any change in the sensation itself, or in the external qualities suggested by it. Habit, for instance, will never cause a person to mistake gentian or quassia for sugar, but it may induce an appetite or liking for what is bitter, and a disgust for what is sweet. No person perhaps was originally delighted with the taste of opium or tobacco, they must at first be highly disgusting to most people; but custom not only reconciles the taste to them, but they become grateful, and even necessary.

Almost every species of sensation becomes grateful or otherwise, according to the force of the impression; for there is no sensation so pleasant, but, that, by increasing its intensity, it will become ungrateful, and at length intolerable. And, on the contrary, there are many which on account of their force are naturally unpleasant, but become, when diminished, highly pleasant. The softest and sweetest sounds may be increased to such a degree as to be extremely unpleasant: and when we are in the steeple of a church, the noise of a peal of bells stuns and confounds our senses, while at a distance their effect is very pleasant. The smell of musk likewise at a distance, and in small quantity, is pleasant; but when brought near, or in large quantity, it becomes highly disagreable. The same may be observed with respect to the objects of the other senses.

For a similar reason, many sensations which are at first pleasing, cease to delight by frequent repetition; though the impression remains the same. This is so well known that illustrations are unnecessary. Those who are economical of their pleasures, or who wish them to be permanent, must not repeat them too frequently. In music, a constant repetition of the sweetest and fullest chords, cloys the ear; while a judicious mixture of them with tones less harmonious will be long relished. Those who are best acquainted with the human heart need not be told, that this observation is not confined to music.

On the same principle likewise we can account for the pleasure afforded by objects that are new; and why variety is the source of so many pleasures; why we gradually wish for an increase in the force of the impression in proportion to its continuance.

The pleasures of the senses are confined within narrow limits, and can neither be much increased nor too often repeated, without being destructive of themselves; thus we are admonished by nature, that our constitutions were not formed to bear the continual pleasures of sense; for the too free use of any of them, is not only destructive of itself, but induces those painful and languid sensations so often complained of by the voluptuary, and which not unfrequently produce a state of mind that prompts to suicide.

As the transition from pleasure to pain is natural, so the remission of pain, particularly if it is great, becomes a source of pleasure. There is much truth, therefore, in the beautiful allegory of Socrates, who tells us, that Pleasure and Pain were sisters, who, however, met with a very different reception by mankind on their visit to the earth; the former being universally courted, while the latter was carefully avoided: on this account, Pain petitioned Jupiter, who decreed that they should not be parted; and that whoever embraced the one, obtained also the other.

There is a great diversity with respect to the duration of the pleasures of the different senses: some of the senses become soon fatigued, and lose the power of distinguishing accurately their different objects: others, on the contrary, remain perfect a long time. Thus smell and taste are soon satiated; hearing more slowly; while, of all the external senses, the objects of sight please us the longest. We may, however, prolong the pleasures of sense by varying them properly, and by a proper mixture of objects or circumstances which are indifferent, and afford less delight. But the very constitution of our nature limits our enjoyments, and points out the impossibility of perpetual pleasures in this state of our existence. To a person who is thirsty, water is delicious nectar; to one who is hungry, every kind of food is agreeable, and even its smell pleasant; to a person who is hot and feverish, the cool air is highly refreshing. But to the same persons in different circumstances, the same things are not only indifferent, but even disgusting; for instance, a person cannot bear the sight or smell of food, after having satiated himself with it, and perpetual feasting will cloy the appetite of the keenest epicure.

I shall conclude this account of the general laws of sensation, by a short recapitulation of those laws.

And, in the first place, it may be observed, that the energy or force of any sensation, is proportioned to the degree of attention given by the mind to the external object which causes it.

Secondly, A repetition of sensations diminishes their energy, and at last nearly destroys it; but this energy is restored by rest, or the absence of these sensations.

Thirdly, The mind cannot attend to two impressions at the same time: so that two sensations never act with the same force at the same instant; the stronger generally overcoming the weaker. The mind, however, can attend to the weaker sensation, in such a manner, as to overpower the stronger, or to render it insensible.

Having fully considered the general laws of sensation, I shall now proceed to examine those proper to each sense; and in this examination, two objects will engage our attention. 1. The structure of the organ which receives and transmits the impulse to the mind. 2. The qualities or properties of external bodies, particularly those by which they are fitted to excite sensation.

The first sense that we shall examine is touch, which, of all the external senses, is the most simple, as well as the most generally diffused. By means of this sense, we are capable of perceiving various qualities and properties of bodies, such as hardness, softness, roughness, smoothness, temperature, magnitude, figure, distance, pressure, and weight; this sense is seldom depraved; because the bodies, whose properties are examined by it, are applied immediately to the extremities of the nerves, without the intervention of any medium liable to be deranged, as is the case with the eye, and ear.

The organ of touch is seated chiefly in the skin, but different parts of this covering possess different degrees of sensibility. The skin consists of three parts. 1. The cutis vera, or true skin, which covers the greatest part of the surface of the body. When the skin is examined by a microscope, we find it composed of an infinite number of papillae, or small eminencies, which seem to be the extremities of nerves, each of which is accompanied by an artery and a vein, so that when the vessels of the skin are injected, the whole appears red. 2. Immediately over the true skin, and filling up its various inequalities, lies a mucous reticulated substance, which has been called by

Malpighi, who first described it, rete mucosum The real skin is white in the inhabitants of every climate; but the rete mucosum is of various colours, being white in Europeans, olive in Asiatics, black in Africans, and copper coloured in Americans. This variety depends chiefly on the degree of light and heat; for, if we were to take a globe, and paint a portion of it with the colour of the inhabitants of corresponding latitudes, we should have an uniform gradation of shade, deepening from the pole to the equator.

The diversity of colour depends upon the bleaching power of the oxygen, which, in temperate climates, combines more completely with the carbonaceous matter deposited in the rete mucosum; while, in hotter climates, the oxygen is kept in a gaseous state by the heat and light, and has less tendency to unite with the carbonaceous matter. In proof of this, the skins of Africans may be rendered white by exposure to the oxygneated muriatic acid.

Over the rete mucosum is spread a fine transparent membrane, called the cuticle, or scarf skin, which defends the organ of feeling from the action of the air, and other things which would irritate it too powerfully. In some parts of the body this membrane is very thick, as in the soles of the feet, and palms of the hands; and this thickness is much increased by use and pressure.

In general, the thinner the cuticle is, the more acute is the sense of touch. This sense is very acute and delicate about the ends of the fingers, where we have the most need of it; but in the lips, mouth, and tongue, it is still more delicate; a galvanic or electrical shock being felt by the tongue, when it is impossible for us to perceive it by the fingers.

This sense, like the others, becomes more exquisite when its organ is defended from the action of external bodies; it is on this account that the cuticle becomes so sensible under the end of the nail, which defends it from the action of external objects; and when part of the nail is taken away, we can scarcely bear to touch any thing with this newly exposed part of the skin.

When we place our fingers upon the surface of any body, the first sensation we experience is that of resistance, after which the other properties are perceived in a natural order; such as heat or cold, moisture or dryness, motion or rest, distance, and figure or shape.

With respect to the diseases of this sense, it is very seldom that it becomes too acute over the whole body; though it frequently does so in particular parts, which may arise from the cuticle being too thin or abraded, or from an inflamed state of the part.

It however becomes sometimes obtuse, and indeed almost abolished over the whole body; and this takes place from compression of the brain, and various affections of the nervous power. This diminution is called anaesthesia. The touch becomes deficient, and indeed almost abolished, when the cuticle is injured by the frequent application of hot bodies, or acrid substances: thus the cuticle of the hands of blacksmiths and glassblowers is generally so hard and horny, that they can take up and grasp in their hand pieces of redhot iron with impunity.

We generally refer pain to this sense, though it may arise from too violent an impression made upon any of the organs of sense.

Pain is an unpleasant sensation, which the mind refers to some part of the body, and very accurately, if any part of the surface is affected, but less so, if it arises from the affection of an internal part. The sensation of pain may arise from any thing which tends to injure the structure of the body, whether that be internal or external; so that it serves as a monitor to put us on our guard, and to induce us to remove any thing which might be injurious to us. This sensation is produced by any thing which punctures, cuts, tears, distends, compresses, bruises, corrodes, burns, or violently stimulates any part of the body.

A moderate degree of pain in any part excites the action of the whole body; a greater quantity of blood and nervous energy is determined to the part. A still greater degree of pain brings on inflammation and its consequences, and if it be intense, it will bring on fever, convulsions, delirium, fainting, and even death.

The endurance of pain depends much on the strength of mind possessed by the patient, which, in some instances, is such, that the most violent pains are patiently endured; while in other instances, the slightest can scarcely be born.

It is a curious circumstance, that a moderate degree of pain, when unaccompanied by fever, often tends to render the understanding more clear, lively, and active. This is confirmed by the experience of people labouring under gout. We have an account of a man who possessed very ordinary powers of understanding, but who exhibited the strongest marks of intelligence and genius in consequence of a severe blow on the head; but that he lost these powers when he recovered from the effects of the blow. Pechlin mentions a young man, who during a complaint originating from worms, possessed an astonishing memory and lively imagination, both of which he nearly lost by being cured. Haller mentions a man who was able to see in the night, while his eyes were inflamed, but lost this power as he got well. All these facts show, that a certain action or energy is necessary for the performance of any of the functions of the body or mind; and whatever increases this action will, within certain limits, increase those functions.

Feeling is by far the most useful, extensive, and important of the senses, and may be said indeed to be the basis of them all. Vision would be of very little use to us, if it were not aided by the sense of feeling; we shall afterwards see that the same observation may be applied to the other senses. In short, it is to this sense that we are indebted, either immediately or indirectly, for by far the greatest part of our knowledge; for without it we should not be able to procure any idea with respect to the magnitude, distance, shape, heat, hardness or softness, asperity or smoothness of bodies; indeed, if we were deprived of this sense, it is difficult to say whether we should have any idea of the existence of any external bodies; on the contrary, it seems probable that we should not.

LECTURE VI. TASTE AND SMELL.

From the sense of touch we proceed naturally to that of taste, for there seems to be less difference between these two senses than between any of the others. The sense of taste appears to be seated chiefly in the tongue; for any sweet substance, such as sugar, applied to any other part of the mouth, scarcely excites the least sensation of taste. The same may be observed with respect to any other sapid body, which, unless it is strongly acrid or irritating, produces no effect on any other part than the tongue; but if it is possessed of much acrimony, it then not only affects the palate, and uvula, but even the oesophagus.

The tongue is a muscular substance, placed in the mouth, connected by one end with the adjacent bones and cartilages, while the other end remains free, and easily moveable. The tongue is furnished, particularly on its upper surface, with innumerable nervous papillae, which are much larger than those I described as belonging to the skin. These papillae are of a conical figure, and extremely sensible, forming, without doubt, the true organ of taste; other papillae are found between them, which are partly conical, and partly cylindrical.

Over the papillae of the tongue is spread a single mucous, and semipellucid covering, which adheres firmly to them, and serves the purpose of a cuticle.

Under these papillae are spread the muscles which make up the fleshy part of the tongue: these are extremely numerous, and by their means the tongue possesses the power of performing a great variety of motions with surprising velocity.

The arteries leading to the tongue are extremely numerous; and, when injected with a red fluid, the whole substance appears of a beautiful red.

The tongue is likewise furnished with a large supply of nerves, some of which undoubtedly serve to supply its muscles with nervous energy, while others terminate in the papillae, and form the proper organ of taste: this office seems to be performed by the third branch of the fifth pair of nerves. The papillae, before described, are formed or composed of a number of small nerves, arteries, and veins, firmly united together by cellular substance. These papillae are excited to action by the application of any sapid body; in consequence of which they receive a greater supply of blood, become enlarged, and vastly more sensible.

The structure of the tongue differs in different animals, which likewise possess corresponding differences with respect to taste. In those quadrupeds, in which it is armed with sharp points, the sense of taste is by no means acute. The same is the case with birds and reptiles, whose tongues are very dry and rough.

In a former lecture I took notice of a liquor which is secreted by the glands of the mouth and neighbouring parts, which is called saliva. This liquor acts an important part in the production of taste; it does not differ much from water, excepting by containing a quantity of mucilage; and nothing is sapid, or capable of affecting the sense of taste, unless it is in some degree soluble in this liquor. Hence earthy substances, which are nearly insoluble, have little or no taste.

It is not, however, sufficient that the substance be possessed of solubility alone; it is necessary likewise that it should be possessed of saline properties, or, at least, of a kind of acrimony, which renders it capable of stimulating the nervous papillae. Hence it is that those substances which are less saline, and less acrid than the saliva, have no taste.

We are capable of distinguishing various kinds of taste, but some of them with less accuracy than others. Among the different kinds of taste, the following have been considered by Haller, and some other physiologists, as primitive: sweet, sour, bitter, and saline. The others have been thought to be compounded of these; for the sense of taste, as well as sight and hearing, is capable of perceiving compound impressions. To these primitive tastes, Boerhaave added alkaline, spirituous, aromatic, and some others. Of these, in different proportions, all the varieties of tastes, which are extremely numerous, are composed.

Some tastes are pleasant and agreeable, others disagreeable, and scarcely tolerable: there is, however, a great diversity in this respect experienced by different persons; for the same taste, which is highly grateful to some, is extremely unpleasant to others.

But the most pleasant tastes, agreeably to the general laws of sensation, which I described in the last lecture, become gradually less pleasant, and at last disgusting; while, on the contrary, the most disagreeable savours, such as tobacco, opium, and assafoetida, become, by custom, not only tolerable, but highly agreeable.

Nature designed this difference of tastes that we might know and distinguish such foods as are salutary; for we may in general observe, that no kind of food which is healthy, and affords proper nutriment to the body, is disagreeable to the taste; nor are any that are ill tasted proper for our nourishment. Those substances, therefore, which possess strong or disagreeable savours, and which, in general, possess a power of producing great changes on our constitution, are to be ranked as medicines, and only to be used when the constitution is deranged; whereas, in general, those which are pleasant, or mild tasted, are proper for nourishing the body. We are therefore excited or prompted to receive nourishment by the pleasant smell or taste of the food; but the avidity with which we take it depends much on the state of the stomach, and likewise on a certain inanition or emptiness; for the coarsest food is grateful to those who are hungry, and whose digestion is good; whereas, to those who have lately eaten, or whose digestive powers are impaired, the most delicate food affords little pleasure. While we are eating, the saliva flows into the mouth more copiously, which excites a more acute sensation of taste. This flow of saliva is likewise frequently excited by the smell or sight of substances agreeable to the taste, which causes an appetite, or desire of eating, similar to that caused by an accumulation of gastric juice in the stomach.

In brute animals, who have not, like ourselves, the advantage of learning from each other by instruction, the faculty of taste is much more acute, by which they are admonished to abstain from noxious or unhealthy food. This sense, for the same reason, is more acute in savages than in those who live in civiilsed society, which, whatever perfection it gives to the reasoning faculties of man, certainly diminishes the acuteness of all our senses, partly by affording fewer inducements to exercise them, and partly by our manner of living, and by the application of substances to the organs of sense, which tend to vitiate them, and render them depraved.

Taste is modified by age, temperament, habit, and disease; and in this it obeys the general laws of sensation. Children are pleased with the taste of what is sweet, and little stimulating; as we advance in years the taste of more stimulating substances becomes agreeable to us; so that we are admonished by this sense to take into the stomach the kind of nourishment fitted to each period of life. We often, however, counteract this salutary monitor by depraving our sense of taste, by the too free use of vinous or spirituous liquors, which so far deadens the sense of taste, that sweet substances become unpleasant, and nothing but acrid and stimulating things can make an impression on our diminished and vitiated sense of taste.

This sense, as well as others, is liable to be diseased. In order that the sense may be perfect, it is necessary that the membrane which envelopes the nervous papillae of the tongue, and serves as a cuticle, should neither be too thick nor too thin, too dry nor too moist. It is necessary likewise that the qualities of the saliva be natural; for alterations in the nature of this liquor affect very much the sense of taste; if it is bitter, which sometimes happens in bilious complaints, all kinds of food have a bitter taste; if it is sweet, the food has a faint and unpleasant flavour; and if it is acid, the food too tastes sour.

This sense is seldom observed to be too acute, unless from a vitiated state of the cuticle, or membrane, which covers the tongue: if this has been abraded or ulcerated, then the substances applied to the tongue are more sensibly tasted; in many instances, however, they do not produce an increased sensation of taste, but only of pain.

The sense of taste, as well as of touch, may become deficient, from various affections of the brain and nerves; this, however, is not often the case. Some persons have naturally a diminished sense of taste, and this generally accompanies a diminished sense of smell. This sense is frequently diminished in sensibility from a deficiency of saliva, as well as of the proper moisture of the tongue. Hence, in many diseases, it becomes defective, such as fevers, colds, and the like; both from a want of the proper degree of moisture, and from defect of appetite, which, as was before observed, is necessary to the sense of taste.

The sense of taste is often diminished by a thickened mucous covering of the tongue, which prevents the application of substances to its nervous papillae. This mucous covering arises from a disordered state of the stomach, as well as from several other affections of the body: hence physicians inspect the tongue that they may be able to judge of the general state of the body; and next to the pulse, it is undoubtedly the best criterion that we have, as it not only points out the nature and degree of several fevers, but likewise, in many instances, the degree of danger to be apprehended.

Having examined the sense of taste, I shall now proceed to consider that of smell; the use of which, like taste, is to enable us to distinguish unwholesome from salutary food; indeed, by this sense, we are taught to avoid what is prejudicial before it reaches the sense of taste, to which it might be very injurious; and thus we are enabled to avoid any thing which has a putrid tendency, which, if received into the stomach, would taint the whole mass of fluids, and bring on speedy dissolution.

The seat of this sense is a soft pulpy membrane, full of pores, and small vessels, which lines the whole internal cavity of the nose. On this membrane are distributed abundance of soft nerves, which arise chiefly from an expansion of the first pair of nerves coming from the brain. This membrane is likewise plentifully supplied with arteries; so that by means of this nervous and arterial apparatus, this membrane is possessed of very great sensibility; but the nerves of the nose being almost naked, require a defence from the air, which is continually drawn through the nostrils into the lungs, and forced out again by respiration. Nature has therefore supplied this part with a thick insipid mucus, very fluid at its first separation, but gradually thickening, as it combines with oxygen, into a dry crust, approaching often to a membranous matter. This mucus is poured out, or exhaled, by the numerous minute arteries of the nostrils, and serves to keep the nervous apparatus moist, and in a proper state for receiving impressions, as well as to prevent the violent effects which might arise from the stimulus of the air and other bodies. The sense of smell is the most acute about the middle of the septum of the nose, where the nervous membrane which I have described is thicker and softer, than in the cavities more deeply situated, where it is less nervous and vascular. These parts are not however destitute of the sense.

As taste proceeds from the action of the soluble parts of bodies on the nervous papillae of the tongue, so smell is occasioned by minute and volatile particles flying off from bodies, which become mixed with the air, and drawn up with it into the nostrils, where these small particles stimulate or act upon the nerves before described, and produce the sensation which we call smelling.

The air therefore, being loaded with the subtile and invisible effluvia of bodies, is, by the powers of respiration, drawn through the nose, so as to apply these particles to the almost naked olfactory nerves, which, as was before observed, excites the sense of smelling.

When we wish to smell accurately, we shut the mouth, open the nostrils as wide as possible, and making a strong inhalation, draw up a greater number of these volatile particles, than could be drawn up by the common action of respiration, by which means the olfactory nerves are more stimulated, and produce a stronger sensation.

In order that this sense may be enjoyed in perfection, it is necessary that the organ of smell be in a proper state or condition to receive impressions, and that the odorous bodies be likewise in a proper state. With respect to the first, it is necessary that the state of the nerves be sound, and particularly that they be kept in a proper state with respect to moisture.

With regard to the odorous bodies, it is necessary, first, that their minute particles should be disengaged, either by heat, friction, fermentation, or other means capable of decomposing those bodies which are the subjects of smell: secondly, that they may be capable of assuming the vaporous or gaseous state, by combining with caloric, or at any rate, that they should remain for a certain time dissolved or suspended in the air: thirdly, that they should not meet with any substance in their way to the nostrils, which is capable of neutralising them, or altering their properties by its chemical action.

Notwithstanding all the pains which physiologists have taken to detect the nature of odorous bodies, they have met with but little success. They are so extremely minute as to escape the other senses, and we can only say that they must be composed of particles in an extreme state of division and subtilty, because very small quantities of odorous matter exhale a sufficient quantity of particles to fill a large space. A grain of camphor, musk, or amber exhales an odour which penetrates every part of a large apartment, and which remains for a long time.

There is perhaps no substance in nature which is absolutely incapable of being changed from a solid state into that of a fluid or gas, by combining with caloric; though different substances require very different quantities of heat to produce those effects. Those which are with difficulty converted into fluids or gases, are termed fixed, while those which are easily changed are called volatile; though these are only terms of comparison, for there is probably no body which is absolutely fixed, or incapable of being reduced to vapour by the application of a sufficient degree of heat.

The odorous property is probably as general as that of being convertible into gas. There is perhaps no body so hard, compact, and apparently inodorous, as to be absolutely incapable of exciting smell by proper methods: two pieces of flint rubbed together, produce a very perceptible smell. Metals which appear nearly inodorous, excite a sensation of smell by friction, particularly lead, tin, iron, and copper. Even gold, antimony, bismuth, and arsenic, under particular circumstances, give out peculiar and powerful odours. The odour of arsenic in its metallic state, and in a state of vapour, resembles that of garlic. The chief means of developing the odorous principles are friction, heat, electricity, fermentation, solution, and mixture. The effect of mixture is very remarkable in the case of lime and muriate of ammoniac, neither of which, before mixture, has any perceptible odour.

There is perhaps then no body which is perfectly inodorous, or entirely destitute of smell: for those which have been generally accounted such, may be rendered odorous by some of the methods I have mentioned.

Several naturalists and physiologists, such as Haller, Linneus, and Lorri, have attempted to reduce the different kinds of odours to classes, but without any great success; for we are by no means so well acquainted with the physical nature of the odorous particles, as we are with that of light, sound, and the objects of touch; and till we do obtain a knowledge of these circumstances, which perhaps we never shall, it will be in vain to attempt any accurate classification. The division of them into odours peculiar to the different kingdoms, is very inaccurate; for the odour of musk, which is thought to be peculiarly an animal odour, is developed in the solution of gold by some mineral solvents; it is perceptible in the leaves of the geranium moschatum, and some other vegetables. The smell of garlic is possessed by many vegetables, by arsenic, and by toads. The violet smell is perceived in some salts, and in the urine of persons who have taken turpentine. The same may be observed with respect to several other odours.

As taste keeps guard, or watches over the passage by which food enters the body, so smell is placed as a sentinel at the entrance of the air passage, and prevents any thing noxious from being received into the lungs by this passage, which is always open. Besides, by this sense, we are invited or induced, to eat salutary food, and to avoid such as is corrupted, putrid, or rancid. The influence of the sense of smell on the animal machine is still more extensive: when a substance which powerfully affects the olfactory nerves is applied to the nostrils, it excites, in a wonderful manner, the whole nervous system, and produces greater effects in an instant, than the most powerful cordials or stimulants received by the mouth would produce in a considerable space of time. Hence in syncope or fainting, in order to restore the action of the body, we apply volatile alkali, or other strong odorous substances, to the nostrils, and with the greatest effect. It may indeed for some time supply the place, and produce the effects, of solid nutriment usually received into the stomach We are told that Democritus supported his expiring life, and retarded, for three days, the hour of death, by inhaling the smell of hot bread, when he could not take nutriment by the stomach. Bacon likewise gives us an account of a man who lived a considerable time without meat or drink, and who appeared to be nourished by the odour of different plants, among which were garlic, onions, and others which had a powerful smell. In short, the stimulus which active and pleasant odours give to the nerves, seems to animate the whole frame; and to increase all the senses, internal and external.

The perfection of the organ of smell is different in different animals; some possessing it very acutely; others on the contrary having scarcely any sense of smell. We may in general observe that this sense is much more acute in many quadrupeds than in man: an in them the organ is much more extensive: in man, from the shape of the head, little opportunity is given for extending this organ, without greatly disfiguring the face. In the dog, the horse, and many other quadrupeds, the upper jaw being large, and full of cavities, much more extension is given to the membrane which is the organ of smell, which in some animals is beautifully plaited, in order to give it more surface. Hence a dog is capable of following game, or of tracing his master in a crowd, or in a road where it could not be done by the mere track. Nay, we are told of a pickpocket being discovered in a crowd, by a dog who was seeking its master, and who was directed to the man by the pocket handkerchief of his master, which the pickpocket had stolen. In dogs the sense of smell must be uncommonly delicate, to enable them to distinguish the way their master has gone in a crowded city.

The habit of living in society, however, deadens this sense in man as well as taste; for we have the advantage of learning the properties of bodies from each other by instruction, and have therefore less occasion to exercise this sense; and the less any sense is exercised, the less acute will it become; hence it is, that those whom necessity does not oblige to to exercise their senses and mental faculties, and who have nothing to do but lounge about, and consume the fruits of the earth, become half blind, half deaf, and, in general, have great deficiency in the sense of smell. The use of spirituous liquors, and particularly of tobacco in the form of snuff, serves likewise in a remarkable manner to deaden this sense.

Savages, however, who are continually obliged to exercise all their senses, have this, as well as others, in very great perfection. Their smell is so delicate and perfect, that it approaches to that of dogs. Soemmering and Blumenbach indeed assert, that in Africans and Americans the nostrils are more extended, and the cavities in the bones lined with the olfactory membrane much larger than in Europeans.

I have already observed the powerful effects which some odours have upon the nervous system. There are some which agreeably excite it, and produce a pleasant and active state of the mind, while others, on the contrary, produce the most terrible convulsions, and even fainting. Those particular antipathies with respect to smells, arise sometimes from something in the original constitution of the body, with which we are unacquainted, but generally from the senses having been powerfully and unpleasantly affected by certain odours at an early period of life. The latter may often be cured by resolution and perseverance, but the former cannot.

The sense of smell sometimes becomes too acute, either from a vitiated state of the organ itself, which is not often the case; or from an increased sensibility or irritability of the whole nervous system, which is observed in hysteria, phrenitis, and some fevers.

This sense is however more often found deficient; and this may arise from a fault in the brain or nerves, which may either proceed from external violence, or from internal causes. A defect of smell often arises from a vitiated state of the organ itself; for instance, if the nervous membrane is too dry, or covered with a thick mucus; of both of which we have an example in catarrh or common cold, where, at the beginning, the nostrils feel unusually dry, but as the disease advances, the pituitary membrane becomes covered with a thick mucus: in both states, the sense of smell is in general deficient, and sometimes nearly abolished.

This sense is sometimes depraved, and smells are perceived when no odorous substance is present; or odours are perceived to arise from substances, which are very different from those which we perceive in a sound state.

There are many diseases likewise of the nose, and neighbouring parts, which cause a depraved sensation; such as ulcers, cancer, caries; a diseased state of the mouth, teeth, throat, or lungs; or a vitiated state of the stomach, which sometimes exhales a vapour similar to that of sulphureted hydrogen. This sense likewise sometimes becomes depraved from a diseased state of the brain and nerves.

LECTURE VII. SOUND AND HEARING.

Having in the last lecture examined the senses of taste and smell, I now proceed to that of hearing. As the sense of smell enables us to distinguish the small particles of matter which fly off from the surfaces of bodies, and float in the air, so that of hearing makes us acquainted with the elastic tremors or impulses of the air itself.

The sense of hearing opens to us a wide field of pleasure, and though it is less extensive in its range than that of sight, yet it frequently surmounts obstacles that are impervious to the eye, and communicates information of the utmost importance, which would otherwise escape from and be lost to the mind.

Sound arises from a vibratory or tremulous motion produced by a stroke on a sounding body, which motion that body communicates to the surrounding medium, which carries the impression forwards to the ear, and there produces its sensation. In other words, sound is the sensation arising from the impression made by a sonorous body upon the air or some other medium, and carried along by either fluid to the ear.

Three things are necessary to the production of sound; first, a sonorous body to give the impression; secondly, a medium or vehicle to convey this impression; thirdly, an organ of sense or ear to perceive it. Each of these I shall separately examine.

Strictly speaking, sonorous bodies are those whose sounds are distinct, of some duration, and which may be compared with each other, such as those of a bell or a musical string, and not such as give a confused noise, like that made by a stone falling on the pavement. To be sonorous, a body must be elastic, so that the tremors exerted by it in the air may be continued for some time: it must be a body whose parts are capable of a vibratory motion when forcibly struck.

All hard bodies, when struck return more or less of a sound; but those which are destitute of elasticity, give no repetition of the sound; the noise is at once produced and dies; while other bodies, which are more elastic and capable of vibration, repeat the sounds produced several times successively. These last are said to have a tone; the others are not allowed to have any. If we wish to give nonelastic bodies a tone, it will be necessary to make them continue their sound, by repeating our blows quickly upon them. This will effectually give them a tone; and an unmusical instrument has often by this means a fine effect in concerts. The effects of a drum depend upon this principle. Gold, silver, copper, and iron, which are elastic metals, are sonorous; but lead, which possesses scarcely any elasticity, produces little or no tone. Tin, which in itself has very little more sound than lead, highly improves the tone of copper when mixed with it. Bell metal is formed of ten parts of copper, and one of tin. Each of these is ductile when separate, though tin is only so in a small degree, yet they form when united a substance almost as brittle as glass, and highly elastic. So curious is the power of tin in this respect, that even the vapour of it, when in fusion, will give brittleness to gold and silver, the most ductile of all metals. Sonorous bodies may be divided into three classes; first, bells of various figures and magnitudes: of these such as are formed of glass have the most pure and elegant tones, glass being very elastic, and its sound very powerful; secondly, pipes of wood or metal; thirdly, strings formed either of metallic or animal substances. The sounds given by strings are more grave or more acute according to the thickness, length, and tension of the strings.

Air is universally allowed to be the ordinary medium of sound, or the medium by which sounds are propagated from sonorous bodies, and communicated to the ear. This may be shown by an experiment with the air pump; also with the condenser.

But though air is the general vehicle of sound, yet sound will go where no air can convey it; thus the scratching of a pin at the end of a long piece of timber may be heard by an ear applied at the other end, though it could not be heard at the same distance through the air. On this account it is that sentinels are accustomed to lay their ears to the ground, by which means they can often discover the approach of cavalry, at a much greater distance than they can see them.

For the same reason two stones being struck together under water, may be heard at a much greater distance by an ear placed under water likewise, than it can be heard through the air. Dr. Franklin, who several times made this experiment, thinks that he has heard it at a greater distance than a mile. This shows that water is better adapted to convey sound than air.

When an elastic body is struck, that body, or some part of it, is made to vibrate. This is evident to sense in the string of a violin or harpsichord, for we may perceive by the eye, or feel by the hand, the trembling of the strings, when by striking they are made to sound. If a bell be struck by a clapper on the inside, the bell is made to vibrate. The base, of the bell, is a circle, but it has been found that by striking any part of this circle on the inside, that part flies out, so that the diameter which passes through this part of the base will be longer than the other diameter. The base, by the stroke, is changed into an ellipse or oval, whose longer axis passes through the part against which the clapper is struck. The elasticity of the bell restores the figure of the base, and makes that part which was forced out of its place, return back to its former situation, from which the same principle throws it out again; so that the circular figure of the bell will be again changed to an ellipse, only now the shorter axis will pass through the part which was first struck.

The same stroke, which makes the bell vibrate, occasions the sound, and as the vibrations decay, the sound grows weaker. We may be convinced by our senses that the parts of the bell are in a vibratory motion while it sounds. If we lay the hand gently on it, we shall easily feel this tremulous motion, and even be able to stop it, or if small pieces of paper be put upon the bell, its vibrations will put them in motion.

These vibrations in the sounding body will cause undulations or waves in the air; and, as the motions of one fluid may often be illustrated by those of another, the invisible motions of the air have been properly enough compared to the visible waves of water produced by throwing a stone therein. These waves spread themselves in all directions in concentric circles, whose common centre is the spot where the stone fell, and when they strike against a bank or other obstacle, they return in the contrary direction to the place from whence they proceeded. Sound in like manner expands in every direction, and the extent of its progress is in proportion to the impulse on the vibrating chord or bell.

Such is the yielding nature of fluids, that when other waves are generated near the first waves, and others again near these, they will perform their vibrations among each other without interruption; those that are coming back will pass by those that are going forwards, or even through them, without interruption: for instance, if we throw a stone into a pond, and immediately after that, another, and then a third, we shall perceive that their respective circles will proceed without interruption, and strike the banks in regular succession.

The atmosphere in the same manner possesses the faculty of conveying sounds in the most rapid succession or combination, as distinctly as they were produced. It possesses the power not only of receiving and propagating simple and compound vibrations in direct lines from the voice, or an instrument, but of retaining and repeating sounds with equal fidelity after repeated reflection and reverberation, as is evident from the sound of a French horn among hills.

Newton was the first who attempted to demonstrate that the waves or pulses of the air are propagated in all directions round a sounding body, and that during their progress and regress they are twice accelerated and twice retarded, according to the law of a pendulum vibrating in a cycloid. These propositions are the foundation of almost all our reasoning concerning sound. When sonorous bodies are struck, they, by their vibration, excite waves in the air, similar to those caused by a stone thrown into water; some parts of these waves entering the ear, produce in us that sensation which we call sound. How these pulsations act upon the auditory nerve, to produce sound, we know not, as we see no necessary connexion between the pulses and the sensation, nor the least resemblance between them; but we can trace their progress to a certain extent, which I shall now endeavour to do.

The external part of the ear is called the auricle, or outward ear, which is a cartilaginous funnel, connected to the bones of the temple, by means of cellular substance, and likewise by its own proper ligaments and muscles. This cartilage is of a very compound figure, being a kind of oval, marked with spirals standing up, and hollows interposed, to which other hollows and ridges correspond on the opposite side. The outer eminence is called helix. Within the body of the cartilage arises a forked eminence called antihelix, which terminates in a small and short tongue called antitragus. The remaining part of the ear, called the concha or shell, is anteriorly hollow, but posteriorly convex, growing gradually deeper; with a crooked line or ridge running along its middle, which is immediately joined to the meatus auditorius, or entrance into the ear; before which stands a round moveable appendix, which serves as a defense, called tragus.

Against this funnel of the ear the sonorous waves strike, and its different parts are most admirably contrived to reflect them all into the meatus auditorius: if it would not occupy too much time, it might be shown, that all these curves and spirals are contrived in the best manner possible, and with a most perfect knowledge of the geometry of sounds, to reflect the sonorous pulses accurately, and in the greatest possible quantity, into the ear.

This external part of the ear is differently formed in different animals; and admirably suited to their various situations and habits. In man it is close to the head, but so formed as to collect the various pulses with great accuracy; in other animals it is more simple, where less accuracy is required, but it is, in general, much larger, having the appearance of an oblong funnel; and this gives them a greater delicacy of hearing, which was necessary for them.

In animals which are defenceless and timid, and which are constantly obliged to seek their safety in flight, the opening of this funnel is placed behind, that they may better hear the noises behind them. This is particularly instanced in the hare. Beasts of prey have this opening before, that they may more easily discover their prey; as the lion and tiger. Those that feed on birds have the opening directed upwards, as the fox; and it is inclined downwards in animals, such as the weasel, which seek their prey on the earth.

To this external part of the ear, which I have described, is connected the meatus auditorius, or passage to the internal ear, which is somewhat of a compressed cylindrical figure, lessening as it bends inwards: a considerable part of it is bony, and it is bent towards the middle. Across this passage, at its inner extremity, is stretched a thin membrane, called membrana tympani. Upon the surface of this membrane, the sonorous waves, which have been directed inwards by the external ear, strike, and cause it to vibrate like the membrane of a drum. This membrane is stretched over a cavity in the bone, called the os petrosum, which cavity is called the tympanum, or drum of the ear, which is of a rounded figure, divided in its middle by a promontory, and continued backwards to the cells of the mastoid bone. Besides this continuation of the tympanum into the mastoid cells, it has a free communication with the mouth, by means of a tube I shall soon describe.

Within this cavity of the tympanum are placed four small bones, which facilitate the hearing: the first is the malleus or hammer, so called from its shape: the upper part of its round head rests upon the concavity of the tympanum, from whence the handle is extended down, along the membrane of the tympanum; this bone has several muscles, which move it in different directions, and cause it to stretch or brace the membrana tympani, when we wish to hear with accuracy.

Connected with the malleus is another small bone, called the incus, or anvil, which is connected with another, called the stapes, or stirrup, from its shape. These two bones are connected by a small oval shaped bone, called os orbiculare, placed between them: the whole forming a little chain of bones.

The stapes, or stirrup, has its end of an oval shape, which fits a small hole called fenestra ovalis, in that part of the ear called the labyrinth, or innermost chamber of the ear.

The labyrinth consists of three parts; first, the vestibule, which is a round cavity in a hard part of the os petrosum; secondly, the semicircular canals, so called from their shape, which however is not exactly semicircular; thirdly, the cochlea, which is a beautifully convoluted canal, like the shell of a snail. This part has a round cavity called fenestra rotunda, which is covered with a thin elastic membrane, and looks into the tympanum.

The vestibule, semicircular canals, and cochlea, the whole of which is called the labyrinth, form one cavity, which is filled with a very limpid fluid resembling water, and the whole lined with a fine delicate membrane, upon which the auditory nerve is expanded, like the retina upon the vitreous humor of the eye. This beautiful apparatus was only lately discovered by an Italian physician, Scarpa. The auditory nerve is a portion of the seventh pair, which is called the portio mollis or soft portion.

There is one part of the ear still to be described, namely, the Eustachian tube, so called from Eustachius, the anatomist, who first described it. This tube opens by a wide elliptical aperture into the tympanum behind the membrane; the other end, which gradually grows wider, opens into the cavity of the mouth. By this canal the inspired air enters the tympanum to be changed and renewed, it likewise serves some important purpose in hearing, with the nature of which we are yet unacquainted. It is certain that we can hear through this passage, for if a watch be put into the mouth, and the ears stopped, its ticking may be distinctly heard; and in several instances of deafness, this tube has been found completely blocked up.

The waves, which have been described as propagated in the air, in all directions from the sounding body, enter the external cartilaginous part of the ear, which, as has before been observed, is admirably fitted for collecting and condensing them. As soon as these pulses excite tremors in the membrane of the tympanum, its muscles stretch and brace it, whence it becomes more powerfully affected by these impulses. It is on this account that we hear sounds more distinctly when we attend to them, the membrane being then stretched.

A tremulous motion, being excited in this membrane, is communicated to the malleus annexed to it, which communicates it to the incus, by which it is propagated through the os orbiculare to the stapes, which imparts this tremulous motion through the foramen ovale to the fluid contained in the labyrinth. This tremor is impressed by the waves excited in this fluid, on every part of the auditory nerve in the labyrinth. The use of the foramen rotundum, or round hole, before described, is probably the same as that of the hole in the side of a drum; it allows the fluid in the labyrinth to be compressed, otherwise it could not vibrate.

If the organization is sound, and tremors are communicated to the auditory nerve, they are in some way or other conveyed to the mind, but in what manner we cannot tell. Nature has hid the machinery by which she connects material and immaterial things entirely from our view, and if we try to investigate them, we are soon bewildered in the regions of hypothesis.

Tremors may however be communicated to the auditory nerve in a different manner from what I have described. If a watch be put between the teeth, and the ear stopped, tremors will be communicated to the teeth, by them to the bones of the upper jaw, and by these to the auditory nerve. In this way a person born deaf, and having no power of hearing through the medium of the air, may become sensible of the pleasures of music.

That sound may be propagated by vibrations, independent of pulses of the air, is evident from the experiment with the string and poker.

There is, strictly speaking, no such thing existing as sound; it being only a sensation of the mind, caused by tremors of the air, or vibrations of the sounding body.

In order to understand more clearly how pulses, or waves are caused by the vibration of bodies, and the manner in which vibrating bodies are affected, I shall just enumerate some of the properties of pendulums, which however I shall not stop to demonstrate here, as that would consume much time.

When two pendulums vibrate which are exactly of the same length, their vibrations are performed in equal times; if they set out together to describe equal arcs, they will agree together in their motions, and the vibrations will be performed in equal times.

But if one of these pendulums be four times as long as the other, the vibrations of the longer will be twice as slow as those of the shorter; the number of vibrations being as the square roots of their lengths.

A pendulum is fixed to one point, but a musical string is extended between two points, and in its vibrations may be compared to a double pendulum vibrating in a very small arc, hence we see how strings of different lengths may agree in their motions after the manner of pendulums; but we must observe that it is not necessary to quadruple the length of a musical string, in order to make the time of vibration twice as long; it will be sufficient merely to double it. We know that from whatever height a pendulum falls on one side, to the same height will it rise on the other. In the same manner will an elastic string continue to vibrate from one side to the other for some time, till its motion be destroyed by the resistance of the air, and friction about its fixed points, and each of its small vibrations, like those of a pendulum, will, for the same reason, be performed in times exactly equal to each other.

Thus we gain from the analogy between a pendulum and a musical string, a more adequate conception of a subject which was never understood till this analogy was discovered. It explains to us why every musical string preserves the same pitch from the beginning to the end of its vibration; or as long as it can be distinguished by the ear; and why the pitch remains still unvaried whether the sound is loud or soft, and all this because the vibrations of the same pendulum whether they are longer or shorter, when compared among themselves, are found to be all performed in equal times till the pendulum be at rest, the difference of the space, which is moved over, compensating for the slowness of the motion till its decay.

To illustrate this subject still further, suppose we have a piece of catgut stretched between two pins; I lay hold of it in the middle and pull it sideways; I let it go, and you will observe that it first straightens itself or returns to its original position. This depends on the elasticity of its particles, which tend to reunite when they have been separated by an external force, just in the same way that the particles of a piece of caoutchouc or Indian rubber attract each other when pulled asunder; and this force not only enables the string to restore itself to its former situation, but will carry it nearly to an equal distance on the other side, just in the same manner as a ball falling down an inclined plane will rise nearly to the same height up another, or a pendulum will rise nearly to the height from which it fell.

In this way will a string move backwards and forwards, till friction and the resistance of the air have destroyed the velocity which it acquired by the force of elasticity.

It is obvious that when a string is thus let fly from the finger, whatever be its own motion, such will also be the motion of the particles of the air which fly before it: the air will be driven forwards, and by that means condensed. When this condensed air expands itself, it will expand not only towards the string, but as its elasticity acts in all directions, it will also expand itself forwards and condense the air that is beyond it, this last condensed air, by its expansion, will produce the same effect on the air that lies still further forwards, and thus the motion produced in the air, by the vibration of the elastic string, is constantly carried forwards and conveyed to the ear.

It will be proper however to observe, that these pulses are sometimes produced without any such vibration of the sounding body, as we find it in musical strings and bells. In these cases we have to discover by what cause these condensations or pulses may be produced without any apparent vibrations in what is considered as the sounding body. We have two or three instances of this kind; one in wind instruments, such as the flute or organ pipe; another in the discharge of a gun.

In an organ, or flute, the air, which is driven through the pipe, strikes against the edge of the lips of the instrument in its passage, and by being accumulated there, is condensed, and this condensation produces waves or pulses in the air.

When a gun is discharged, a great quantity of air is produced, by the firing of the gunpowder, which being violently propelled from the piece, condenses the air that encompasses the space where the expansion happens; for whatever is driven out from the space where the expansion is made will be forcibly driven into the space all around it. This condensation forms the first pulse, and as this, by its elasticity, expands again, pulses of the same sort will be produced and propagated forwards.

There is likewise another curious instance of the production of sound, when a tube is held over a stream of inflamed hydrogen gas issuing out of a capillary tube in a bottle.

Sounding bodies propagate their motions on all sides, directly forwards, by successive condensations and rarefactions, so that sound is driven in all directions, backwards and forwards, upwards and downwards, and on every side; the pulses go on succeeding each other like circles in disturbed water.

Sounds differ from each other both with respect to their tone and intensity: in respect to their tone, they are distinguished into grave and acute: in respect to their intensity, they are distinguished into loud and low, or strong and weak. The tone of a sound depends on the velocity with which the vibrations are performed, for the greater the number of vibrations in a given time, the more acute will be the tone, and on the contrary, the smaller the number, the more grave it will be. The tone of a sound is not altered by the distance of the ear from the sounding body; but the intensity or strength of any sound depends on the force with which the waves of the air strike the ear; and this force is different at different distances; so that a sound which is very loud when we are near the body that produces it, will be weaker if we are further from it, though its tone will suffer no alteration; and the distance may be so great that we cannot hear it at all. It has been demonstrated, that the intensity of sound at different distances from the sounding body is inversely as the square of the distance.

Sound moves with the same velocity at all distances from the sounding body, otherwise it would not produce the same tone at all distances. Sounds of different tones likewise move with the same velocity. This is evident from a peal of bells being heard in the same order in which they are rung, whether we are near, or at a distance.

It is likewise found that sounds of the same tone but of different intensities are propagated with the same velocity. A low sound cannot indeed be heard so far as a loud one; but sounds, whether low or loud, will be conveyed in an equal time to any equal distance at which they can both be heard. The report of a cannon does not move faster, or pass over a given space sooner, than the sound of a musical string.

The principal cause of the decay of sound is the want of perfect elasticity in the air: whence it happens that every subsequent particle has not the entire motion of the preceding particle communicated to it, as is the case with equal and perfectly elastic bodies; consequently the further the motion is propagated, the more will the velocity with which the particles move be diminished; the condensation of the air will be diminished also, and consequently its effect on the ear. That the want of perfect elasticity in the air is the principal cause of the decay of sound, appears from this, that sounds are perceived more distinctly when the north and easterly winds prevail, at which time the air is dry and dense, as appears from the hygrometer and barometer; and, of course, the air in this state must be more elastic, for the vapours diffused through the atmosphere, unless dilated by intense heat, diminish the spring of the air.

That sound is not propagated to all distances instantaneously, but requires a sensible time for its passage from one place to another, is evident from the discharge of a gun at a distance; for the report is not heard till some time after the flash is seen. Light moves much more swiftly than sound; it comes from the sun in eight minutes, which is at the rate of 74,420 leagues in a second; so that the velocity of light may be considered as instantaneous, at any distance on the earth; and, as sound takes up a considerable time in its passage, the interval between the flash and the report of the gun shows the space it passes over in a given time, which is found to be 1142 feet in a second; so that if three seconds elapse between the time when we see the flash and hear the report of the gun, it must be distant 1142 yards.

From experiments that have been made at different times, by various philosophers, we may collect the following results. First, That the mean velocity of sound is a mile in about 4 3/4 seconds, or 1142 feet in a second of time. Secondly, That all sounds, whether they be weak or strong, have the same velocity. Thirdly, That sound moves over equal spaces in equal times, from the beginning to the end.

The tone of a musical string, or a bell, appears continuous. This depends upon a law of sensation, formerly mentioned, namely, that impressions made upon any of the organs of sense do not immediately vanish, but remain some time; and we hear sound continuous from these vibrations, for the same reason that we hear it continuous when we draw a stick quickly along a rail, or a quill along the teeth of a comb; the vibrations succeed each other so quickly that we hear the succeeding before the effect of the preceding is worn off; though it must be evident that the impression produced by each pulse or wave of the air is perfectly distinct and insulated.

The act of combining sounds in such a manner as to be agreeable to the ear, is called music. This art is usually divided into melody and harmony. An agreeable succession of sounds is called melody; but when two or more sounds are produced together, and afford an agreeable sensation, the effect is called harmony. When two sounds are produced together, and afford pleasure to the sense of hearing, the effect is called a concord; but when the sensation produced is harsh or disagreeable, it is called a discord. These different effects seem to depend upon the coincidence of the vibrations of the two strings, and consequently on the coincidence of the pulses which they excite in the air. When the strings are equally stretched, and of the same length and thickness, their vibrations will always coincide, and they produce a sound so similar to each other, that it is called unison, which is the most perfect concord. When one string is only half the length of the other, the vibrations coincide at every second vibration of the shorter string: this produces a compound sound, which is more agreeable to the ear than any other, except the unison; this note, when compared with the tone produced by the longer string, is called the octave to it, because the interval between the two notes is so divided by musicians that from one to the other they reckon eight different tones.

If the strings be of the length, two and three, the coincidence of the pulses will happen less frequently, viz. at every third vibration of the shorter string, and the concord will be less perfect. This forms what is called a fifth. The less frequent the coincidence of the vibrations, the less perfect will be the concord, or the less pleasure will it afford to the mind; till the vibrations coincide so seldom, that the sound produced by both strings at once is harsh and disagreeable, and is called a discord.

The effects of music upon the mind, the power by which it moves the heart, touches the passions, and excites sometimes the highest pleasure, and sometimes the deepest melancholy, depend upon melody. By a simple melody the ignorant are affected as well as those skilled in music. The pleasures arising from harmony or a combination of sounds are acquired rather than natural. Its pleasures are the result of experience and knowledge in music; music affords a source of innocent and inexhaustible pleasure, but its effects are different on different persons: some are enthusiastically fond of it, while others hear the sweetest airs with a listlessness bordering upon indifference. This has been supposed to depend on a musical ear, which is not given by nature to all. The cause of this difference is by no means evident. It does not depend on the delicacy of the sense of hearing, for there are some persons half deaf, who have the greatest relish for music; while others who have a very acute sense of hearing have no relish for music. In some instances I think a musical ear has been acquired where it did not seem originally to exist.

The force of sound is increased by the reflection of many bodies, particularly such as are hard or elastic, which receive the waves or pulses of the air and reflect them back again; these reflected pulses, striking the ear along with the original, strengthen the original sound. Hence it is, that the voice of a speaker is louder, and more distinctly heard, in a room than in the open air. I said that these reflected sounds entered the ear at the same time with the original:

this however is not strictly the case, for they must enter the ear after the original, because the sound has a greater space to move over: but they enter the ear so quickly after the original that our sense cannot distinguish the difference. If however the reflecting body should be placed at such a distance, that the reflected sound should enter the ear some considerable or sensible time after the original, an echo or distinct sound would be heard.

It appears from experiment that the ear of an experienced musician can only distinguish such sounds as follow each other at the rate of nine or ten in a second, or any lower rate; and therefore that we may have a distinct perception of the direct and reflected sound, there should at least be an interval of 1/9 of a second; but in this time sound passes over one hundred and twenty seven feet, and consequently, unless the space between the sounding body and the reflecting surface, added to that between the reflecting surface and the ear, be greater than one hundred and twenty seven feet, no echo will be heard, because the reflected sound will enter the ear so soon after the original, that the difference cannot be distinguished; and therefore it will only serve to augment the original sound.

From what has been said, it is evident, in order that a person may hear the echo of his own voice, that he should stand at least sixty three, or sixty four feet from the reflecting obstacle, so that the sound may have time to move over at least one hundred and twenty seven feet before it come to his ear, otherwise he could not distinguish it from the original sound.

But though the first reflected pulses may produce no echo, both on account of their being too few in number, and too rapid in their return to the ear; yet it must be evident that the reflecting surface may be so formed, that the pulses, which come to the ear after two or more reflections, may, after having passed over one hundred and twenty seven feet or more, arrive at the ear in sufficient numbers to produce an echo, though the distance of the reflecting surface from the ear be less than the limit of echoes. This is instanced by the echoes that we hear in several caves or caverns.

The sense of hearing is more apt to be vitiated or diseased than any of the other senses, which indeed is not surprising, when we consider that its organ is complex, consisting of many minute parts, which are apt to be deranged.

It sometimes becomes too acute, and this may arise either from too great an irritability of the whole nervous system, which often occurs in hysteria, also in phrenitis, and some fevers; or from an inflamed state of the ear itself.

The sense of hearing becomes diminished, and often entirely abolished; and this may arise from various causes, such as an original defect in the external ear, or the meatus auditorius, or both; the meatus auditorius is often blocked up with wax or other substances, which being removed, the hearing becomes perfect. Deafness may likewise arise from a rigidity of the membrane of the tympanum, from its being erodedor ruptured, or from an obstruction of the Eustachian tube. It may likewise arise from a paralysis or torpor of the auditory nerve, or from some diseased state of the labyrinth, or from a vitiated state of the brain and nerves. There is a kind of nervous deafness which comes on suddenly, and often leaves the patient as suddenly.

There are various instances, however, in which the membane of the tympanum has been lacerated or destroyed, without a total loss of the sense of hearing, or indeed any great diminution of it. A consideration of these circumstances induced Mr. Astley Cooper to think of perforating it, in cases of deafness arising from a permanent obstruction of the Eustachian tube, and he has often performed this operation with great success. Of this he has given an account in the last part of the Philosophical Transactions. This operation ought however only to be performed in case of the closure of the Eustachian tube. Cases of this kind may be distinguished by the followingcriteria. If a person on blowing the nose violently, feel a swelling in the ear, from the membrane of the tympanum being forced outwards, the tube is open; and though the tube be closed, if the beating of a watch placed between the teeth, or pressed against the side of the head, cannot be heard, the operation cannot relieve, as the sensibility of the auditory nerve must have been destroyed. In a closed Eustachian tube, there is no noise in the head, like that accompanying nervous deafness.

There is one species of deafness, which occurs very frequently, and happens generally to old persons, though sometimes to the delicate and irritable in the earlier periods of life Anxiety and distress of mind have been known to produce it. Its approach is generally gradual, the patient hears better at one time than at another; a cloudy day, a warm room, agitated spirits, or the operation of fear, will produce a considerable diminution in the powers of the organ. In the open air the hearing is better than in a confined situation; in a noisy, than in a quiet society; in a coach when it is in motion, than when it is still. A pulsation is often felt in the ear; a noise resembling sometimes the roaring of the sea, and at others the ringing of distant bells is heard. This deafness generally begins with a diminished secretion of the wax of the ear, which the patient attributes to cold. It may be cured, particularly at its commencement, by the application of such stimulants as are capable of exciting a discharge from the ceruminous glands; for which purpose they should be introduced into the meatus auditorius.

In some cases of this kind, where the auditory nerve has been in some degree torpid, or rather perhaps where there has been a kind of paralysis, or want of action, in the muscles which brace the membrane of the tympanum, and keep the chain of bones in their proper state; a person has not been able to hear, except during a considerable noise. Willis mentions the case of a person who could only hear when a drum was beaten near her; and we are told of a woman who could not hear a word except when the sound of a drum was near, in which case she could hear perfectly well. When she married, her husband hired a drummer for his servant. In instances of this kind the noise probably excites the action of the torpid muscles, which then put the apparatus in a proper condition to hear.

LECTURE VIII. VISION.

In order to understand properly the theory of vision, it will be necessary to premise an anatomical description of the eye: but I shall content myself with as short a one as will suffice to explain the effects it produces on the rays of light, so as to produce the distinct vision of an object.

The shape of the eye is nearly spherical; it is composed of several coats or tunics, one within another; and is filled with transparent humours of different densities.

The proper coats of the eye are reckoned five in number; viz. the sclerotica, cornea, choroides, iris or uvea, and the retina.

After the tunica conjunctiva, or adnata, (a membrane, which, having lined the eyelids in the manner of a cuticle, surrounds the anterior part of the globe) is removed, we perceive a white, firm, membrane, called the sclerotica, which takes its rise from that part of the globe where the optic nerve enters, and surrounds the whole eye, except a little in the fore part; which fore part has a membrane, immediately to be described, called the cornea. The tunica sclerotica, viewed through the conjunctiva, forms what is called the white of the eye. Some anatomists have supposed that this coat is a continuation of the dura mater, which surrounds the optic nerve; but later observations have shown this opinion to be ill founded. The tunica sclerotica consists of two layers, which are with difficulty separated.

The next coat is the cornea, so called from its resemblance to transparent horn; it arises where the sclerotic coat ends, and forms the fore part of the eye. The cornea is a segment of a lesser sphere than the rest of the eye, and consequently makes it more prominent on the fore part: it is transparent, and firmly connected by its edges to the sclerotica.

Immediately adherent to the sclerotica, within, is the choroides, which takes its rise from that part of the eye where the optic nerve enters, and accompanies the sclerotica to the place where it is joined to the cornea; here it is very closely connected to the sclerotica, where it forms that annulus, called ligamentum ciliare; then leaving the sclerotic coat, it is turned inwards, and surrounds the crystalline lens; but as this circle, where it embraces the crystalline, is much narrower than where the membrane leaves the sclerotic coat, it becomes beautifully corrugated, which folds or corrugations have been, by the more ancient anatomists, improperly called ciliary processes.

To the same part of the choroid coat, where the ciliary ligament begins, is fixed a moveable and curious membrane, called the iris; this membrane has a perforation in the middle, called the pupil, for the admission of the rays of light. The iris is composed of two kinds of fibres: those of the one sort tend, like the radii of a circle, towards its centre, and the others form a number of concentric circles round the same centre. The pupil is of no constant magnitude, for when a very luminous object is viewed, the circular fibres of the iris contract, and diminish its orifice; and, on the contrary, when objects are dark and obscure, those fibres relax, and suffer the pupil to enlarge, in order to admit a greater quantity of light into the eye: it is thought that the radial fibres also assist in enlarging the pupil. The iris is variously coloured in different persons, but according to no certain rule; though in general, they who have light hair, and a fair complexion, have the iris blue or grey; and, on the contrary, they whose hair and complexion are dark, have the iris of a deep brown; but whether this difference in colour occasions any difference in the sense, is not yet discovered. In the human eye the whole choroid coat, and even the interior surface of the iris or uvea, is lined with a black mucus; this mucus, or as it is called, pigmentum, is darkest in young persons, and becomes more light coloured as we advance in years. In many animals, but more particularly those which catch their prey in the night, this pigmentum is of a bright colour: its use will appear afterwards.

The last, and innermost coat of the eye, is the retina, it differs much from the above mentioned coats, being very delicate and tender. It is nothing but an expansion of the medullary part of the optic nerve, which is inserted into each eye, nearer the nose, and a little higher, than the axis. This coat has been thought by many to end where the choroides, going inwards, towards the axis of the eye, forms the ciliary ligament; Dr. Monro thinks that it is not continued so far, and we cannot see with what advantage it could have been continued to the ciliary ligament, since none of the rays of light, passing through the pupil, could fall upon that part of it. In the middle of the optic nerve is found the branch of an artery, from the internal carotid, which is diffused and ramified in a beautiful manner along the retina. From this artery, a small branch goes through the middle of the vitreous humour, and giving off branches on every side, expands itself upon the capsule of the crystalline lens.

We shall now consider the humours of the eye, which are three in number, the aqueous, the crystalline, and the vitreous; all transparent, and in general colourless; but of different densities.

The aqueous humour, so called from its resemblance to water, fills up all the space between the cornea and the crystalline humour. It is partly before and partly behind the uvea, and is divided by that membrane into two parts, which are called the chambers of the aqueous humour; which chambers communicate with each other by means of the pupil.

The next humour is the crystalline; it is situated between the aqueous and vitreous humours, and is connected to the choroid coat by the ciliary ligament: it is not the least of all the humours, as has been generally supposed, the aqueous and it being of equal weights; but its substance is more firm and solid than that of the other humours: its figure is that of a double convex lens; but the fore part next the pupil is not so convex as its other side, which is contiguous to the vitreous humour; the diameter of the sphere, of which its anterior segment is a part, being in general about seven or eight lines, whereas the diameter of the sphere, of which its posterior segment forms a portion, is commonly only about five or six lines. It is covered with a fine transparent capsule, which is called arachnoides. This humour is situated exactly behind the pupil, but not in the centre of the eye, as was supposed by Vesalius, being a good deal nearer its forepart. The convexity of its posterior surface is received into an equal concavity of the vitreous humour. It is not of an equal density throughout, but is much more hard and dense towards its centre than externally, the reason of which will appear hereafter. Till we arrive at about our thirtieth year, this humour continues perfectly transparent, and colourless; about that time it generally has a little tinge of yellow, and this colour increases with age.

The third humour of the eye, is the vitreous; it is the largest of all the humours, filling up the whole of that part of the eye which lies behind the crystalline humour. It is thicker than the aqueous, but thinner than the crystalline humour; on its back part is spread the retina, and in the middle of its fore part is a small cavity, in which the whole posterior surface of the crystalline lens lies; this humour is also enclosed in a very fine capsule, called tunica vitrea; this capsule at the edge of the crystalline humour is divided into two membranes, of which the one is continued over the whole anterior surface of the vitreous humour, and lines that cavity into which the back part of the crystalline is received; the other passes over the crystalline humour, and covers all its fore part, by which means these two humours are closely connected together. The weights of the aqueous, crystalline, and vitreous humours in a human eye, are, according to the accurate Petit, at a medium, to each other, as 1, 1, and 25.

It was thought necessary to premise this general description of the structure of the eye, in order that what we are going to add in the remaining part of this Lecture may be the more easily comprehended. A more distinct idea will perhaps be had from a contemplation of the following figure, which represents the section of an eye by a vertical plane passing through its centre.

[FIGURE]

EXPLANATION.

NOO represents the optic nerve.
The outmost line ALLB represents the sclerotic coat, and the part ACB the transparent cornea.
The line ALLB, immediately within the former, represents the choroides; the part APB is the iris or uvea, in which the hole at P is the pupil.
The line FOOG is the retina.
The cavity ACBEMDA is the aqueous humour.
DE is the crystalline lens or humour.
The space DFOOGE, lying behind the crystalline, represents the vitreous humour.
BE and AD is the ligamentum ciliare.

Nature and Properties of Light.

After this short description of the human eye, I shall next proceed to take notice of some of the properties of light; but shall confine myself to such as are absolutely necessary for explaining the phenomena of vision, as far as that can be done from optical principles.

1. It is, I believe, generally at present agreed, that light consists of exceedingly small particles of matter, projected with great velocity in all directions from the luminous or radiant body. This hypothesis, to which no solid objection has yet been made, appears to be more simple than any other; and is so consistent with all the phenomena yet observed, that we have great reason to think it true: however, as it is not absolutely and directly demonstrated, it may have been wrong in optical writers to have given this hypothesis (for it can only be called a hypothesis) as a definition of light.

2. The space through which light passes is, by opticians, called a medium, and it is observed, that, when light passes through a medium, either absolutely void, or containing matter of an uniform density, and of the same kind, it always proceeds in straight lines.

3. Those rays of light which come directly from a luminous body to the eye, only give us a perception of light; but when they fall upon other bodies, and are from them reflected to the eye, they give us an idea or perception of those bodies.

4. When a ray of light passes out of one medium into another of different density, it is bent out of its course, and is said to be refracted. We must, however, except those rays which fall in a direction perpendicular to the surface of the refracting medium; as the refractive force acts in the same direction in which those rays move, they will not be turned out of their course, but proceed in the same direction they had before they entered the refracting medium. When a ray passes out of a rarer into a denser medium, it will be refracted, or bent towards a line which is perpendicular to the surface which separates the media at the point where it falls; but when it passes out of a denser into a rarer medium, it will be bent from the perpendicular.

5. Whenever the rays, which come from all the points of any object, meet again in so many points, after they have been made to converge by refraction, there they will form the picture of the object, distinct, and of the same colours, but inverted. This is beautifully demonstrated by a common optical instrument, the camera obscura. If a double convex lens, be placed in the hole of a window shutter in

a dark room, and a sheet of white paper be placed at a certain distance behind the lens; a beautiful, but inverted picture of the external objects will be formed: but if the paper be held nearer, or more remote than this distance, so that the rays from each point shall not meet at the paper, but betwixt it and the lens, or beyond the paper, the picture will be indistinct and confused.

Of the Manner in which Vision is performed.

From the just mentioned properties of light, and the description we have given of the eye, it will not be difficult to explain the theory of vision, so far as it depends upon optical principles. For the eye may, with great propriety, be compared to a camera obscura; the rays which flow from external objects, and enter the eye, painting an inverted picture of those objects on the retina: if you carefully dissect from the bottom of an eye, newly taken out of the head of an animal, a small portion of the tunica sclerotica and choroides, and place this eye in a hole made in the window shutter of a dark chamber, so that the bottom of the eye may be towards you; the pictures or images of external objects will be painted on the retina in lively colours, but inverted.

In order to see how the several parts of the eye contribute to produce this effect, let us follow the rays proceeding from a luminous point, and see what will happen to them from the beforementioned properties of light.

Since the rays of light flow from every visible point of a body in every direction, some of them, issuing from this point, will fall upon the cornea, and, entering a medium of greater density, will be refracted towards the perpendicular, and as they fall upon a convex spherical surface, nearly in a parallel state, the pupil being so extremely small, it is evident, from the principles of optics, that they will be made to converge: those which fall very obliquely will either be reflected, or falling upon the uvea, or pigmentum nigrum, which covers the ciliary ligaments, will be suffocated, and prevented from entering the internal parts of the eye: those which fall more directly, as was before said, become converging, in which state they fall upon the anterior surface of the crystalline humour, which, having a greater refracting power than the aqueous humour, and its surface being convex, will cause them to converge still more, in which state they will fall upon the posterior surface of the crystalline, or anterior surface of the vitreous humour; which having a less refractive power than the crystalline, they will be refracted from the perpendicular; but, as they fall upon a concave surface, it is evident, from the principles of optics, that they will be made to converge still more: in which state they will go on to the retina, and if the eye is well formed, the refraction of these several humours will be just sufficient to bring them to a point or focus on the retina.

The same thing will happen to rays flowing from every other visible point of the object: the rays which flow from every point will be collected into a corresponding point on the retina, and, consequently, will paint the image of that object inverted; the rays coming from the superior part of any object, being collected on the inferior part of the retina, and vice versa, as is manifest from the principles of optics.

If the rays are accurately, or very nearly, collected into a focus on the retina, distinct vision will be produced; but if they be made to converge to a point before or beyond the retina, the object will be seen indistinctly; this is proved by holding a convex or concave glass before the eye of a good sighted person: in the former case, the rays will be made to converge to a point before they arrive at the retina, and in the latter, to a point beyond it. In these cases, it is plain that the rays which flow from a point in the object, will not form a point, but a circular spot, upon the retina, and these various circles intermixing with other, will render the image very indistinct. This is well illustrated by the camera obscura, where if you hold the paper nearer or more remote than the focal distance of the lens, the picture will be indistinct.

So far then, in the theory of vision, are we led by the principles of optics, and we can with certainty, by their assistance, affirm, that if the eye is sound, and the image of an object distinctly painted upon the retina, it will be seen distinctly, erect, and of its proper colours: so far we can proceed on safe and sure grounds, but if we venture further, we shall find ourselves bewildered in the regions of hypothesis and fancy. The machinery by which nature connects the material and immaterial world is hidden from our view; in most cases we must be satisfied with knowing that there are such connexions, and that these connexions invariably follow each other, without our being able to discover the chain that goes between them. It is to such connexions that we give the name of laws of nature; and when we say that one thing produces another by a law of nature, this signifies no more, than that one thing, which is called the cause, is constantly and invariably followed by another, which we call the effect, and that we know not how they are connected. But there seems a natural propensity in the mind of man, to endeavour to account for every phenomenon that falls under his view, which has given rise to a number of absurd and romantic conjectures in almost every branch of science. From this source has risen the vibration of the fibres of the optic nerve, or the undulation of a subtile ether, or animal spirits, by which attempts have been made to explain the theory of vision; but all of them are absurd and hypothetical.

Kepler was the first who had any distinct notion of the formation of the pictures of objects on the bottom of the eye; this discovery he published about the year 1600. Joannes Baptista Porta had indeed got some rude notion of it prior to the time of Kepler, but as he knew nothing of the refraction made by the humours of the eye, his doctrine was lame and defective, for he imagines that the images are painted on the surface of the crystalline humour.

The disputes concerning the theory of vision had very much divided the ancient philosophers; some of them imagining that vision was caused by the reception of rays into the eye; while a great many others thought it more agreeable to nature, that certain emanations, which they called visual rays, should flow from the eye to the object.

We shall now inquire more particularly how each part of the eye is peculiarly fitted to produce distinct vision. Though the eye is composed of different humours, yet one might have been sufficient to collect the rays into a focus, and form the picture of an object upon the retina. By the experiments of the accurate Dr. Robertson, it appears that there is less difference in the density, as well as in the refracting power of the humours, than has been generally thought: by weighing them in a hydrostatic balance, he found that the specific gravities of the aqueous and vitreous humours were very nearly equal, each being nearly equal to that of water: and that the specific gravity of the crystalline did not exceed the specific gravity of the other humours in a greater proportion than that of about 11 to 10. Hence it would seem to follow, that the crystalline is not of such great use in bringing the rays together, and thereby forming the pictures of objects on the retina, as has been commonly thought by optical writers; for though in shape it resembles a double convex lens, and is, on that account, fitted to make the rays converge; yet, be cause it is situated between two humours nearly of the same refractive power with itself, it will alter the direction of the light but a little. From this, the reason is evident why the sight continues after the operation for the cataract, in which the crystalline is depressed, or extracted, and why a glass of small convexity is sufficient to supply the little refraction wanting, occasioned by the loss of this humour. But without doubt, several important purposes are effected by this construction of the eye; which could not have been attained if it had been composed of one humour only. Some of those purposes seem sufficiently evident to us; for instance, by placing the aqueous humor before the crystalline, and partly before the pupil, and making the cornea convex, a greater quantity of light is made to enter the eye than could otherwise have done without enlarging the size of the pupil; the light will also enter in a less diverging state than it could have done if the pupil had been enlarged, and consequently be more accurately collected to a focus on the retina; for a perfect eye can only collect such rays to a focus on that membrane, as pass through the pupil nearly in a state of parallelism.

Another, and perhaps a principal advantage derived from the different humours in the eye, is, probably, to prevent that confusion arising from colour, which is the consequence of the different degrees of refrangibility of the rays of light. From the experiments of Mr. Dollond, it appears, though contrary to the opinion of Newton, and most other optical writers, that different kinds of matter differ extremely with respect to the divergency of colour produced by equal refractions; so that a lens may be contrived, composed of media of different dispersing powers, which will form the image of any object free of colour; this discovery Mr. Dollond has applied to the improvement of telescopes, with great success. It is by no means improbable, that nature has, for the same purpose, placed the crystalline lens betwen two media of different densities, and, probably, different dispersing powers, so that an achromatic image, free from the prismatic colours, will be formed on the retina. Indeed we find a conjecture of this kind, so long since as Dr. David Gregory's time, he says, in speaking of the imperfection of telescopes, "Quod si ob difficultates physicas, in speculis idoneis torno elaborandis, et poliendis, etiamnum lentibus uti oporteat, fortassis media diversae densitatis ad lentem objectivam componendam adhibere utile foret, ut a natura factum observamus in oculo, ubi crystallinus humor (fere ejusdem cum vitro virtutis ad radios lucis refringendos) aqueo et vitreo (aquae quoad refractionem hand absimilibus) conjungitur, ad imaginem quam distincte fieri poterit, a natura nihil frustra moliente, in oculi fundo depingendam."

In describing the eye, I observed, that the crystalline humour was not every where of the same consistence, being much more hard and dense towards its centre, than externally: in the human eye, it is soft on the edges, and gradually increases in density as it approaches the centre: the reason of this construction is evident, at least we know of one use which it will serve; for, from the principles of optics, it is plain that the rays which fall at a distance from the axis of the crystalline, by reason of their greater obliquity, if the humour were of the same density in all its parts, would be more refracted than those which fall near its axis, so that they would meet at different distances behind the crystalline humour; those which pass towards its extremity, nearer, and those near its axis, at a greater distance, and could not be united at the same point on the retina, which would render vision indistinct; though the indistinctness arising from this cause, is only about the 1/5449 part of that which arises from the different refrangibility of the rays of light, as Sir Isaac Newton has demonstrated. Nature has, however, contrived a remedy for this also, by making the crystalline humour more dense and solid near its centre, that the rays of light which fall near its axis, may have their refraction increased, so as to meet at the same point with those which fall at a distance from its axis.

Of the manner in which the Eye conforms itself in order to see distinctly at different Distances.

It has been much disputed in what manner the eye conforms itself to see distinctly at different distances; for it is evident, that, without some change, the rays which flow from objects at different distances, could not be collected into a focus at the same point, and, consequently, though the eye might see distinctly at one distance, it could not at another.

This subject has given rise to a variety of opinions, but few of them are satisfactory; and though several of them might explain the phenomena of vision, at different distances, yet it is by no means proved that those supposed changes do take place in the eye. I shall content myself with just mentioning the principal opinions on this subject, without engaging in a controversy, which has for a long time employed the ingenuity of philosophers to little purpose.

Some are of opinion, that the whole globe of the eye changes its figure; becoming more oblong when objects are near, and more flat when they are removed to a greater distance; and this change in the figure of the eye is differently explained by different authors; some maintain that it is rendered oblong by the joint contraction of the two oblique muscles: others think that the four straight muscles acting together, compress the sides of the globe, and by this compression, reduce it to an oblong figure, when objects are near; and that, by its natural elasticity, it recovers its former figure when these muscles cease to act. Others again think that when these four straight muscles act together, they render the eye flat by pulling it inwards, and pressing the bottom of it against the fat; and that it is reduced to its former figure, either by the joint contraction of the two oblique muscles, or by the inherent elasticity of its parts, which exerts itself when the muscles cease to act.

That, if such a change should take place in the eye, it would produce distinct vision, will be readily granted; but that such a one does not take place, at least in any of these ways, is, in my opinion, very certain. Dr. Porterfield thinks that the crystalline lens has a motion by means of the ligamentum ciliare, by which the distance between it and the retina is increased or diminished, according to the different distances of objects. The ligamentum ciliare, he says, is an organ, the structure and disposition of which excellently qualify it for changing the situation of the crystalline, and removing it to a greater distance from the retina, when objects are too near for us; for that, when it contracts, it will not only draw the crystalline forwards, but will also compress the vitreous humour, lying behind it, so that it must press upon the crystalline, and push it from the retina. Although this hypothesis will, in a great measure, account for distinct vision at different distances, yet it could only be of use where the rays enter the eye with a certain degree of divergency, while, however we are sure, that in looking at very distant objects which are at different distances from us, the eye undergoes a change. But a sufficient objection to Dr. Porterfield's hypothesis is, that it is by no means proved that the crystalline lens can be moved in the manner he supposes, or that the ligamentum ciliare is possessed of muscular fibres; on the contrary some eminent anatomists deny that they are.

We shall now take a view of the opinion of M. de la Hire, who considered this subject, as well as almost every other relating to vision, with the closest attention; he maintains, that, in order to view objects distinctly at different distances, there is no alteration but in the size of the pupil, which is well known to contract and dilate itself according to the quantity of light flowing from the object we look at, being most contracted in the strongest light, and most dilated when the light is weakest; and consequently will contract when an object is held near the eye, and dilate as it is removed, because in the first case the quantity of light entering the eye is much greater than in the last. That this contraction of the pupil will have the effect of rendering vision distinct, especially when objects are within the furthest limits of distinct vision, will plainly appear, if we consider the cause of indistinct vision. Dr Jurin has shown, that objects may be seen with sufficient distinctness, though the pencils of rays issuing from the points of them do not unite precisely in another point on the retina, but instead thereof, if they form a circle which does not exceed a certain magnitude, distinct vision will be produced; the circle formed by these rays on the retina he calls the circle of dissipation. The pupil will, by contracting, not only diminish the circles of dissipation, and thereby help to produce distinct vision, but will also prevent so great a quantity of light from falling near the circumferences of those circles; and Dr. Jurin has shown, that, if the light on the outer side of the circles of dissipation is diminished, the remainder will scarce affect the sense. In both these ways, the contraction of the pupil has a tendency to diminish the circles of dissipation, and, consequently, to produce distinct vision. This is likewise confirmed by experiment, for when an object is placed so near, that the pupil cannot be so much contracted as is necessary for distinct vision, the same end may be obtained by means of an artificial pupil: for, if a small hole is made in a card, a very near object may be viewed through it with the greatest ease and distinctness. Also, if a person have his back turned towards a window, and hold a book so near his eyes as not to be able to read, if he turn his face to the light, he will find, that he will be able to read it very distinctly; which is owing to the contraction of the pupil by means of the light.

M. Le Roi, a member of the Royal Academy of Montpelier, has attempted to defend the opinion of M. de la Hire, and, indeed, it seems, of all others, the best supported by facts; but perhaps it may not account so well for vision at great distances. It is likewise rendered more probable by viewing the pictures of external objects, formed in a dark chamber, by rays coming through a hole in the window shutter; those pictures will be rendered distinct, by dilating, or contracting the aperture, without the assistance of a lens, accordingly as the object is more or less distant; those who have had the crystalline lens depressed, or extracted, by means of one glass can see objects pretty distinctly at different distances. These, and several other arguments that might be brought, tend to prove that the eye accommodates itself to view objects distinctly at different distances, chiefly by means of the motion of the pupil; and though this does not explain the phenomenon so satisfactorily as we could wish, yet it is certain, that it has a share in it; we are however certain, that, in whatever manner it may be produced, the eye has a power of accommodating itself to view objects distinctly enough at several different distances.

Concerning the Seat of Vision.

No subject has been more canvassed than that concerning what is improperly called the seat of vision. In early times, the crystalline lens was thought to be best qualified for this office; but this substance, though situated in the middle of the eye, which Baptista Porta thought to be the proper centre of observation, had universally given place to the better founded pretensions of the retina: and, from the time of Kepler, few ventured to dispute its claim to that office, till M. Mariotte was led, from some curious circumstances, to think that vision was not performed by the retina, but by the choroid coat. Having often observed in the dissections of men, as well as of brutes, that the optic nerve is not inserted exactly opposite to the pupil, that is, in the place where the picture of the objects upon which we look directly, is made: and that in man it is somewhat higher, and on the side towards the nose, he had the curiosity to examine the reason of this structure, by throwing the image of an object on this part of the eye. In order to do this, he fastened on a dark wall, about the height of his eyes, a small round paper, to serve for a fixed point of sight; and he fastened such another paper on the right hand, at the distance of about two feet, but rather lower than the former, so that light issuing from it, might strike the optic nerve of his right eye, while the left was kept shut. He then placed himself over against the former paper, and drew back by degrees, keeping his right eye fixed, and very steady upon it, and when he had retired about ten feet, he found that the second paper entirely disappeared. This, he says, could not be imputed to the oblique position of the second paper, with respect to his eye, because he could see more remote objects on the same side. This experiment he repeated by varying the distances of the paper and his eye. He also made it with his left eye, while the right eye was kept shut, the second paper being fastened on the left side of the point of sight; so that by the situation of the parts of the eye, it could not be doubted that this defect of vision is in the place where the optic nerve enters, where only the choroides isdeficient.

From this he concludes, that the defect of vision is owing to the want of the choroid coat, and, consequently, that this coat is the proper organ of vision. A variety of other arguments in favour of the choroides occurred to him, particularly he observed that the retina is transparent, which he thought could only enable it to transmit the rays further, and he could not persuade himself that any substance could be considered as being the termination of the pencils, and the proper seat of vision, at which the rays are not stopped in their progress.

Mr. Pequet, in answer to Mariotte's observation, says, that the retina is very imperfectly transparent, resembling oiled paper, or horn: and, besides, that its whiteness demonstrates that it is sufficiently opaque for stopping the rays of light as much as is necessary for vision: whereas, if vision be performed by means of those rays which are transmitted through such a substance as the retina, it must be very indistinct.

Notwithstanding the plausibility of this opinion of M. Mariotte, and the number of celebrated men who joined him in it, I must confess, that none of their arguments, though very ingenious, have been able to make me a convert to that opinion.

If we argue from the analogy of the other senses, in all of which the nerves form the proper seat of sensation, we shall be induced to give judgment in favour of the retina. And this argument from analogy is much strengthened, by considering that the retina is a large nervous apparatus, immediately exposed to the impressions of light; whereas the choroides receives but a slender supply of nerves, and seems no more fitted for the organ of vision than any other part of the body. But facts are not wanting which make still more in favour of the retina. It appears from observations made upon the sea calf and porcupine, that these animals have their optic nerves inserted in the axis of the eye, directly opposite the pupil, which renders it very improbable that the defect in sight, where the optic nerves enter, can be owing to the want of the choroides in that place; for were this true, then in those creatures which have the optic nerves inserted in the axis of the eye, and which by consequence do directly receive on the extremity of the nerve the pictures of objects, all objects would become invisible to which their eyes are turned, because the choroides is wanting in that place where the image falls; but this is contrary to experience.

M. Le Cat, though he strenuously supports Mariotte's opinion, takes notice of a circumstance, which, if he had properly considered it, might have led him to a contrary conclusion: from a beautiful experiment he obtains data, which enable him with considerable accuracy to determine the size of the insensible spot in his eye, which he finds to be about 1/30 or 1/40 of an inch in diameter, and consequently only about 1/5 or 1/6 of the diameter of the optic nerve, that nerve being about 1/6 of an inch in diameter. I find that in my eye likewise, the diameter of the insensible spot is about 1/40 of an inch, or something less. Whence it is evident that vision exists where the choroid coat is not present, and consequently that the choroid coat is not the organ of vision.

It is probably owing to the hardness and callosity of the retina where the nerve enters, that we have this defect of sight, as it has not yet acquired that softness and delicacy which is necessary for receiving such slight impressions as those of the rays of light, and this conjecture is rendered still more probable by an observation of M. Pequet, who tells us, that a bright and luminous object, such as a candle, does not absolutely disappear, but one may see its light, though faint. This not only shows that the defect of sight is not owing to a want of the choroides, but also that the retina is not altogether insensible where the nerve enters. These circumstances, in my opinion, render it certain, that the retina, and not the choroid coat, is the organ of vision.

Of our seeing Objects erect by inverted Images.

Another question concerning vision, which has very much perplexed philosophers, is this; how comes it that we see objects erect, when it is well known that their images or pictures on the retina are inverted? The sagacious Kepler, who first made this discovery, was the first that endeavoured to explain the cause of it.

The reason he gives for our seeing objects erect, is this, that as the rays from different points of an object cross each other before they fall on the retina, we conclude that the impulse we feel upon the lower part of the retina comes from above; and that the impulse we feel from the higher part, comes from below. Des Cartes afterwards gave the same solution of this phenomenon, and illustrated it by the judgment we form of the position of objects which we feel with our arms crossed, or with two sticks that cross each other. But this solution is by no means satisfactory: first, because it supposes our seeing objects erect to be a deduction from reason, drawn from certain premises, whereas it seems to be an immediate perception; and secondly, because all the premises from which this conclusion is supposed to be drawn, are absolutely unknown to far the greater part of mankind, and yet they all see objects erect.

Bishop Berkeley, who justly rejects this solution, gives another, founded on his own principles, in which he is followed by Dr. Smith. This ingenious writer thinks that the ideas of sight are altogether unlike those of touch; and since the notions we have of an object by these different senses, have no similitude, we can learn only by experience how one sense will be affected, by what, in a certain manner, affects the other. Thus, finding from experience, that an object in an erect position, affects the eye in one manner, and that the same object in an inverted position, affects it in another, we learn to judge, by the manner in which the eye is affected, whether the object is erect or inverted. But it is evident that Bishop Berkeley proceeds upon a capital mistake, in supposing that there is no resemblance between the extension, figure, and position, which we see, and that which we perceive by touch. It may be further observed, that Bishop Berkeley's system, with regard to material things, must have made him see this question, in a very different light from that in which it appears to those who do not adopt his system.

In order to give a satisfactory answer to this question, we must first examine some of the laws of nature, which take place in vision; for by these the phenomena of vision must be regulated.

It is now, I believe, pretty well established, as a law of nature, that we see every point of an object in the direction of a right line, which passes from the picture of that point on the retina, through the centre of the eye. This beautiful law is proved by a very copious induction of facts; the facts upon which it is founded are taken from some curious experiments of Scheiner, in his Fundamenta Optices. They are confirmed by Dr. Porterfield, and well illustrated by Dr. Reid. The seeing objects erect by inverted images is a necessary consequence of this law of nature: for from thence it is evident that the point of the object whose picture is lowest on the retina, must be seen in the highest direction from the eye; and that the picture which is on the right side of the retina, must be seen on the left.

Of seeing Objects single with two Eyes.

That we should have two pictures of an object, and yet see it single, has long been looked upon as a curious circumstance by philosophers: and of consequence, many attempts have been made to account for it, few of which, however, are satisfactory.

As it would take up too much time to give a view of all the opinions on this subject, I shall pass over the opinions of Galen, Gassendus, Baptista Porta, Rohault, and others, which do not deserve a serious refutation; and shall content myself with making a few observations on the hypothesis of Bishop Berkeley.

But it seems the most proper way of proceeding, first of all to consider the phenomena of single and double vision, in order, if possible, to discover some general principle to which they lead, and of which they are necessary consequences; and, for the sake of perspicuity, we shall premise the following definition.

When a small object is seen single with both eyes, those points on the two retinas on which the pictures of the object fall, may be called corresponding points: and when the object is seen double, we shall call such points, non-corresponding points.

Now we find that in sound and perfect eyes, when the axes of both are directed to one point, an object placed in that point is seen single; and in this case, the two pictures which show the object single, are painted on the centres of the retinas. Hence, the centres of the two retinas correspond.

Other objects at the same distance from the eyes, as that to which their axes are directed, do also appear single: and in this case, it is evident to those who understand the principles of optics, that the pictures of an object to which the eyes are not directed, but which is at the same distance as that to which they are directed, fall both on the same side of the centre, that is, both to the right, or both to the left, and both at the same distance from the centre. Hence it is plain, that points in the retina, which are similarly situated with respect to the centres, are corresponding points.

An object which is much nearer, or much more distant from the eyes, than that to which their axes are directed, appears double. In this case, it will easily appear, that the pictures of the object which is seen double, do not fall upon points which are similarly situated. From these facts, we are led to the following conclusion, viz. that the points of the two retinas, which are similarly situated with respect to the centres, correspond with each other, and that the points which are dissimilarly situated, do not correspond. The truth of this general conclusion is founded upon a a very full induction, which is all the evidence we can have for a fact of this nature.

The next thing that seems to merit our attention, is, to inquire, whether this correspondence between certain points of the two retinas which is necessary to single vision, is the effect of custom, or an original property of the human eyes.

We have a strong argument in favour of its being an original property, from the habit we get of directing our eyes accurately to an object; we get this habit by finding it necessary for perfect and distinct vision; because thereby, the two images of the object falling upon corresponding points, the eyes assist each other in vision, and the object is seen better by both eyes together, than by one: but when the eyes are not accurately directed, the two images of the object fall upon points which do not correspond, whereby the sight of the one eye disturbs that of the other. Hence it is not unreasonable to conclude, that this correspondence between certain points of the retina is prior to the habits we acquire in vision: and, consequently, natural and original.

We have all acquired the habit of directing our eyes in one particular manner, which causes single vision; now if the Author of Nature had ordained that we should see objects single, only when our eyes are thus directed, there is an obvious reason why all mankind should agree in the habit of directing them in this manner; but, if single vision were the effect of custom, any other habit of directing the eyes would have answered the purpose; we therefore, on this supposition, can give no reason why this particular habit should be so universal.

Bishop Berkeley maintains a contrary opinion, and thinks that our seeing objects single with both eyes, as well as our seeing them erect, by inverted images, depends upon custom. In this he is followed by Dr. Smith, who observes, that the question, why we see objects single with both eyes, is of the same nature with this, why we hear sounds single with both ears; and that the same answer will serve for both; whence he concludes, that as the second of these phenomena is the effect of custom, so also is the first. But I think, that the questions are not so much of the same sort, as that the same answer will serve for both; and, moreover, that our hearing single with both ears is not the effect of custom. No person will doubt that things which are produced by custom, may be undone by disuse, or by a contrary custom. On the other hand, it is a strong argument, that an effect is not owing to custom, but to the constitution of nature, when a contrary custom, long continued, is found neither to change nor weaken it. Now it appears, that from the time we are able to observe the phenomena of single and double vision, custom makes no change in them, every thing which at first appeared double, appearing so still in the same circumstances. Dr. Smith has adduced some facts in favour of his opinion, which, though curious, seem by no means decisive. But in the famous case of the young man couched by Mr. Cheselden, after having had cataracts in both his eyes till his thirteenth year, it appears that he saw objects single from the time he began to see with both eyes. And the three young gentlemen mentioned by Dr. Reid, who had squinted, as far as he could learn, from infancy, as soon as they learned to direct both eyes to an object, saw it single.

In these cases it is evident that the centres of the retina corresponded originally, for Mr. Cheselden's young man had never seen at all before he was couched, and the other three had never been accustomed to direct the axes of both eyes to the same point. These facts render it probable, that this correspondence is not the effect of custom, but of fixed and immutable laws of nature.

With regard to the cause of this correspondence, many theories have been proposed, but as none of them can be looked on in any other light than as probable conjectures, I think it would be to little purpose to notice them. That of the illustrious Newton is the most ingenious of any, and though it has more the appearance of truth than any other, that great man has proposed it under the modest form of a query.

Having given a short account of the principal phenomena of vision, I proceed next to treat of some of the diseases to which this sense is subject, I shall first take notice of the deformity called squinting.

Of Squinting.

Though this is a subject which well deserves our particular attention, yet having spoken of such a variety of subjects in the preceding part of this lecture, I have not time for many observations on this. I shall just mention the principal opinions, concerning the cause of this deformity, and point out that which seems to me most probable. This subject is certainly very worthy the attention of the physician, as it is a case concerning which he may often be consulted, and which it may be sometimes in his power to cure.

A person is said to squint, when the axes of both his eyes are not directed to the same object.

This defect consists in the wrong direction of one of the eyes only. I have never met with an instance in which both eyes had a wrong direction, neither have I seen one accurately described by any author.

The generality of writers on this subject have supposed this defect to proceed from a disease of, or want of proper correspondence in, the muscles of the eyes, which not acting in proper concert with one another, as in persons free from this blemish, are not able to point both eyes to the same object. But this, I think, is very seldom the cause, for when the other eye is shut, the distorted eye can be moved by the action of its muscles, in all possible directions, as freely as that of any other person, which shows that it is not owing to a defect in the muscles, neither is it owing to a want of correspondence in the muscles of both eyes; for when both eyes are open, and the undistorted eye is moved in any direction whatever, the other always accompanies it, and is turned the same way at the same instant of time.

I shall next take notice of the hypothesis of M. de la Hire, who supposes, that in the generality of mankind, that part of the retina which is seated in and about the axis of the eye, is of a more delicate sense and perception than what the rest of the coat is endowed with; and therefore we direct both axes to the same object, chiefly in order to receive the picture on that part of the retina which can best perceive it; but in persons who squint, he conceives the most sensible part of the retina of one eye, not to be placed in the axis, but at some distance from it: and that, therefore, this more sensible part of the retina is turned towards the object, to which the other eye is directed, and thus causes squinting. This ingenious hypothesis has been followed by Dr. Boerhaave, and many other eminent physicians. If it be true, then if the sound eye be shut, and the distorted eye alone be used to look at an object, it must still be as much distorted as before, for the same reason: but the contrary is true in fact; for if you desire such a person to close his other eye and look at you with that which is usually distorted, he will immediately turn the axis of it directly towards you. If you bid him open the undistorted eye, and look at you with both eyes, you will find the axis of this last pointed at you and the other turned away, and drawn close to the nose, or perhaps to the upper eye lid. From these facts, and some others mentioned by Dr. Jurin, I think we may conclude that this defect is seldom, if ever, occasioned by such a preternatural make of the eye, as M. de la Hire supposed.

From the most accurate observations it will appear, that by far the most common cause of squinting, is a defect in the distorted eye. Dr. Reid examined above twenty people that squinted, and found in all of them a defect in the sight of one eye; M. Buffon likewise, from a great number of observations, has found that the true and original cause of this blemish, is an inequality in the goodness, or in the limits of distinct vision, in the two eyes. Dr. Porterfield says this is generally the case with people who squint; and I have found it so in all that I have had an opportunity of examining.

With regard to the nature of this defect, the distorted eye is sometimes more convex, and sometimes more flat, than the sound one; sometimes it does not depend upon the convexity, but upon a weakness, or some other affection, of the retina, of the nature of which we are ignorant.

It will be easy to conceive how this inequality of goodness in the two eyes, when in a certain degree, must necessarily occasion squinting, and that this blemish is not a bad habit, but a necessary one, which the person is obliged to learn, in order to see with advantage. When the eyes are equally good, an object will appear more distinct and clear when viewed with both eyes than with only one; but the difference is very little. The ingenious Dr. Jurin, who has made some beautiful experiments to ascertain this point, has shown, that when the eyes are equal in goodness, we see more distinctly with both than with one, by about one thirteenth part only. But M. Buffon has found that when the eyes are unequal, the case will be quite different. A small degree of inequality will make the object, when seen with the better eye alone, appear equally bright or clear, as when seen with both eyes; a little greater inequality will make the object appear less distinct when seen with both eyes, than when it is seen with the stronger eye alone; and a still greater inequality will render the object, when seen with both eyes, so confused, that in order to see it distinctly, one will be obliged to turn aside the weak eye, and put it into a situation where it cannot disturb the sight of the other. The truth of this may be easily proved by experiment. Let a person take a convex lens, and hold it about half an inch before one of his eyes; he will, by these means, render them very unequal. and if he attempt to read with both eyes, he will perceive a confusion in the letters, caused by this inequality; which confusion will disappear as soon as he shuts the eye which is covered with the lens, and looks only with the other.

Squinting is a necessary consequence of this inequality in the goodness of the two eyes; for a person whose eyes are to a certain degree unequal, finds that, when he looks at an object, he sees it very indistinctly; every conformation, or change of direction of his eyes which lessens the evil, will be agreeable; and he will acquire a habit of turning his eye towards the nose, not for the sake of seeing better with it, but in order to avoid, as much as possible, seeing at all with the distorted eye; for which purpose it will be drawn either under the upper eye lid, so that the pupil may be entirely or partially covered; or directly towards the nose, in which case the image of the object will fall at a distance from the axis of the eye: and it is a fact well known to philosophers, that we never naturally attend to an image which is at a distance from the axis; so that in this situation it will give little disturbance to the sight of the other.

It is easy to show that a squinting person very seldom, if ever, sees an object with the distorted eye. Indeed in above forty cases that I have examined, I found that when I placed an opaque body between the undistorted eye and the object, it immediately disappeared, nor were they able to see it at all, till they directed the axis of the distorted eye to the object. I find the same observation made by Dr. Reid and M. Buffon.

M. Buffon takes notice of a fact which I have often observed; viz. that many persons have their eyes very unequal without squinting.

When the difference is very considerable, the weak eye does not turn aside, because it can see almost nothing, and therefore cannot disturb the vision of the good eye. Also, when the inequality is but small, the weak eye will not turn aside, as it affords very little disturbance to the sight of the other: when the inequality consists in the difference of convexity, or difference of the limits of distinct vision, having the limits of distinct vision in each eye given, it may be calculated with some degree of accuracy what degree of inequality is necessary to produce squinting. It seems then that there are certain limits with regard to the inequality of the eyes, necessary to produce this deformity; and that if the inequality be either greater or less than these limits, the person will not squint.

Having now endeavoured to show what is the most common cause of squinting, I shall briefly attempt to point out those cases in which we may expect to effect a cure, and afterwards give a very short account of the most likely methods of doing it.

We cannot have great hopes of success, when there is a very great defect in the distorted eye. When the eyes are of different convexities, there is no other way of removing the deformity, than by bringing them to an equality by means of glasses, and then the person would only look straight when he used spectacles. When this defect is owing to a weakness in the distorted eye, it may sometimes be cured: M. Buffon observes that a weak eye acquires strength by exercise, and that many persons, whose squinting he had thought to be incurable, on account of the inequality of their eyes, having covered their good eye for a few minutes only, and consequently being obliged to exercise their bad one for that short time, were themselves surprised at the strength it had acquired, and on measuring their view afterwards, he found it to be more extended, and judged the squinting to be curable. In order therefore to judge with any certainty of the possibility of a cure, it ought always to be tried whether the distorted eye will grow better by exercise; if it does not, we can have little hopes of success; but when the eyes do not differ much in goodness, and it is found that the distorted eye acquires strength by exercise, a cure may then be attempted: and the best way of doing it, (according to M. Buffon) is to cover the good eye for some time, for, in this condition, the distorted eye will be obliged to act, and turn itself towards objects, which by degrees will become natural to it.

When the eyes are nearly brought to an equality by exercise, but cannot both be directed to the same point, Dr. Jurin's method may be practised, which is as follows.

If the person is of such an age, as to be capable of observing directions, place him directly before you, and let him close the undistorted eye, and look at you with the other; when you find the axis of this fixed directly upon you, bid him endeavour to keep it in that situation, and open the other eye; you will now see the distorted eye turn away from you towards his nose, and the axis of the other will be pointed towards you, but with patience and repeated trials he will, by degrees, be able to keep the distorted eye fixed upon you, at least for some time after the other is opened, and when you have brought him to keep the axis of both eyes fixed upon you, as you stand directly before him, it will be time to change his posture, and set him, first a little to one side of you, and then to the other, and so practise the same thing. And when, in all these situations, he can perfectly and readily turn the axes of both eyes towards you, the cure is effected. An adult person may practice all this before a mirror, without a director, though not so easily as with one: but the older he is, the more patience will be necessary.

With regard to the success of this method, M. Buffon says, that having communicated his scheme to several persons, and, among others, to M. Bernard de Jussieu, he had the satisfaction to find his opinion confirmed by an experiment of that gentleman, which is related by Mr. Allen, in his Synopsis Universae Medicinae. Dr. Jurin tells us that he had attempted a cure in this manner with flattering hopes of success, but was interrupted by the young gentleman's falling ill of the small pox, of which he died. Dr. Reid likewise tried it with success on three young gentlemen, and had brought them to look straight when they were upon their guard. Upon the whole this seems by much the most rational method of attempting to cure the deformity.

The only remaining morbid affections of the eye which I shall take notice of in this lecture, are two, which produce the indistinct vision of an object, by directly opposite means. The first is caused by the cornea, and crystalline, or either of them, being too convex, or the distance between the retina and crystalline being too great. It is evident, that from any of these causes, or all combined, the distinct picture of an object, at an ordinary distance, will fall before the retina, and therefore the picture on the retina itself must be confused, which will render the vision confused and indistinct; whence, in order to see things distinctly, people whose eyes are so formed are obliged to bring the object very near their eyes; by which means the rays fall upon the eye in a more diverging state, so that a distinct picture will be formed on the retina, by which the object will be distinctly seen: from the circumstance of such persons being obliged to hold objects near their eyes, in order to see them distinctly, they are called short sighted.

If a short sighted person look at an object through a small hole made in a card, he will be able to see even remote objects, with tolerable distinctness, for this lessens the circles of dissipation on the retina, and thus lessens the confusion in the picture. For the same purpose, we commonly observe short sighted people, when they wish to see distant objects more distinctly, almost shut their eye lids: and it is from this, says Dr. Porterfield, that short sighted persons were anciently called myopes.

The sight of myopes is remedied by a concave lens of proper concavity, which, by increasing the divergency of the rays, causes them to be united into a focus on the retina: and they do not require different glasses for different distances, for, if they have a lens which will make them see distinctly at the distance most commonly used by other persons, for example, at the distance at which persons whose eyes are good generally read, they will, by the help of the same glass, be able to see distinctly at all the distances at which good sighted people can see distinctly: for the cause of shortsightedness, is not a want of power to vary the conformation of the eye, but is owing to the whole quantity of refraction being too great for the distance of the retina from the cornea.

The other defect to be mentioned, is of an opposite nature, and persons labouring under it are called long sighted, or presbytae: it is caused by the cornea and crystalline, or either of them, being too flat in proportion to the distance between the crystalline and retina:

whence it follows, that the rays which come from an object at an ordinary distance, will not be sufficiently refracted, and, consequently, will not meet at the retina, but beyond it, which will render the picture on the retina confused, and vision indistinct. Whence, in order to read, such persons are obliged to remove the book to a great distance, which lessens the divergency of the rays falling on the eye, and makes them converge to a focus sooner, so as to paint a distinct image on the retina.

The presbytical eye is remedied by a convex lens of proper convexity, which makes the rays converge to a focus sooner, and thus causes distinct vision: the sight of such persons is even more benefited by a convex lens, than that of myopes by a concave one; for a convex lens not only makes the picture of the object on the retina distinct, but also more bright, by causing a greater quantity of light to enter the pupil; while a concave one, at the same time that it renders vision distinct, diminishes the quantity of light.

Long sighted persons commonly become more so as they advance in years, owing to a waste of the humours of the eye; and even many people whose sight was very good in their youth, cannot see without spectacles when they grow old. The same waste in the humours of the eye, is the reason why shortsighted persons commonly become less so as they advance in years; so that many who were shortsighted in their youth, come to see very distinctly when they grow old. Dr. Smith seems to doubt this, and thinks that it is rather a hypothesis than a matter of fact. I have however myself seen several instances in confirmation of it; and it is very natural to suppose, that since short and long sight depend upon directly opposite causes, and since the latter is increased by age, the former must be diminished by it.

LECTURE IX. THE LAWS OF ANIMAL LIFE.

In the preceding lectures I have taken a view, first of the general structure and functions of the living body, and next of the different organs called senses, by means of which we become acquainted with external objects. I shall next endeavor to show that, through the medium of these different senses, external objects affect us in a still different manner, and by their different action, keep us alive:

for the human body is not an automaton; its life, and its different actions, depend continually on impressions made upon it by external objects. When the action of these ceases, either from their being withdrawn, or from the organization necessary to perceive them, being deranged or injured, the body becomes a piece of dead matter; becomes obedient to the common laws of chemical attraction, and is decomposed into its pristine elements, which, uniting with caloric, form gases; which gases, being carried about in the atmosphere, or dissolved in water, are absorbed by plants, and contribute to their nourishment. These are devoured by animals, which in their turn die, and are decompounded; thus, in the living world, as well as in the inanimate, every thing is subject to change, and to be renewed perpetually.

> "Look nature through, 'tis revolution all,
> All change, no death; day follows night; and night,
> The dying day; stars rise, and set, and rise;
> Earth takes th' example; see the summer gay,
> With her green chaplet, and ambrosial flowers,
> Droops into pallid autumn; winter gray,
> Horrid with frost, and turbulent with storm,
> Blows autumn and his golden fruits away,
> Then melts into the spring; soft spring with breath
> Favonian, from warm chambers of the south
> Recals the first. All to reflourish, fades;
> As in a wheel, all sinks to reascend."

The subject on which we are entering is of the utmost importance; for, by pointing out the manner in which life is supported and modified by the action of external powers, it discovers to us the true and only means of promoting health and longevity, for the action of these powers is generally within our own direction; and if the action of heat, food, air, and exercise, were properly regulated, we should have little to fear from the attacks of diseases.

When we examine the human body, the most curious and unaccountable circumstance that we observe, is its life, or its power of motion, sensation, and thought: for though the structure of the different parts which we have examined must excite our admiration and wonder, each part being admirably fitted for the performance of its different functions, yet without the breath of life, all these beautiful contrivances would have been useless. We have seen that the structure of the eye indicates in its contriver, the most consummate skill in optics; and of the ear the most perfect knowledge of sounds; yet if sensibility had not being given to the nerves which administer to these organs, the pulses of the air might have been communicated to the fluid in the labyrinth, and the rays of light might have formed images in the retina, without our being, in the smallest degree, conscious of their existence.

Though our efforts to discover the nature of life have hitherto been, and perhaps always will be, unsuccessful, yet we can, by a careful induction, or observation of facts, discover the laws by which it is governed, with respect to the action of external objects. This is what I shall now attempt to do.

The first observation which strikes us, is that of the very different effects that are produced when inanimate bodies act on each other, and when they exert their action on living matter.

When dead matter acts upon dead or inanimate matter, the only effects we perceive are mechanical, or chemical; that is, either motion, or the decomposition and new combination of their parts. If one ball strikes another, it communicates to it a certain quantity of motion, this is called mechanical action; and if a quantity of salt, or sugar, be put into water, the particles of salt, or sugar, will separate from each other, and join themselves to the particles of the water; these substances in these instances are said to act chemically on each other, and in all cases whatever, in which inanimate or dead bodies act on each other, the effects produced are motion, or chemical attraction; for though there may appear to be other species of action which sometimes take place, such as electric and magnetic attraction and repulsion, yet these are usually referred to the head of mechanical action or attraction.

But when dead matter acts upon those bodies we call living, the effects produced are much different. There are many animals which pass the winter in a torpid state which has all the appearance of death; and they would continue in that state, if deprived of the influence of heat; now heat if applied to dead matter, will only produce motion, or chemical combination: in fluids it produces motions by occasioning a change in their specific gravity; and we know that it is one of the most powerful agents in chemical combination and decomposition; but these are the only effects it produces when it acts upon dead matter. But let us examine its effects when applied to living organized bodies. Bring a snake or other torpid animal into a moderately warm room, and observe what will be the consequence. After a short time the animal begins to move, to open its eyes and mouth; and when it has been subject to the action of heat for a longer time, it crawls about in search of food, and performs all the functions of life.

Here then, dead matter, when applied to the living body, produces the living functions, sense and motion: for if the heat had not been applied, the animal would have continued senseless, and apparently lifeless.

In more perfect animals, the effects produced by the action of dead matter upon them, are more numerous, and are different in different living systems; but are in general the following; sense and motion in almost all animals, and in many the power of thinking, and other affections of the mind.

The powers, or dead matters, which by their action produce these functions, are chiefly heat, food, and air. The proof that these powers do produce the living functions is in my opinion very satisfactory, for when their action is suspended, the living functions cease. If we take away, for instance, heat, air, and food, from animals, they soon become dead matter. This is as strong a proof that these matters are the cause of the functions, as that heat is the cause of the expansion of bodies, when we find that by withdrawing it the expansion ceases. Indeed it is not necessary that an animal should be deprived of all these powers to put a stop to the living functions; if any one of them is taken away, the body sooner or later becomes dead matter: it is found by experience, that if a man is deprived of air, he dies in about three or four minutes; for instance, if he is immersed under water: if he is deprived of heat, or in other words is exposed to a very severe degree of cold, he likewise soon dies; or if he is deprived of food, his death is equally certain, though more slow; it is sufficiently evident then that the living functions are owing to the action of these external powers upon the body.

What I have here said is not confined to animals, but the living functions of vegetables are likewise caused by the action of dead matter upon them. The powers, which by their actions produce the living functions of vegetables, are principally heat, moisture, light, and air.

From what has been said, it clearly follows, that living bodies must have some property different from dead matter, which renders them capable of being acted on by these external powers, so as to produce the living functions; for if they had not, it is evident that the only effects which these powers could produce, would be mechanical, or chemical.

Though we know not exactly in what this property consists, or in what manner it is acted on, yet we see that when bodies are possessed of it, they become capable of being acted on by external powers, so as to produce the living functions.

We may call this property, with Haller, irritability, or, with Brown, excitability; or we may use vital principle, or any other term, could we find one more appropriate. I shall use the term excitability, as perhaps the least liable to exception, and in using this term, it is necessary to mention that I mean only to express a fact, without the smallest intention of pointing out the nature of that property which distinguishes living from dead matter; and in this we have the illustrious example of Newton, who called that property which causes bodies in certain situations to approach each other, gravitation, without in the least hinting at its nature. Yet though he knew not what gravitation was, he investigated the laws by which bodies were acted on by it, and thus solved a number of phenomena which were before inexplicable: in the same manner, though we are ignorant of the nature of excitability, or of the property which distinguishes living from dead matter, we can investigate the laws by which dead matter acts upon living bodies through this medium. We know not what magnetic attraction is, yet we can investigate its laws: the same may be observed with respect to electricity. If ever we should obtain a knowledge of the nature of this property, it would make no alteration in the laws which we had before discovered.

Before we proceed to the investigation of the laws by which the living principle or excitability is acted on, it will be first necessary to define some terms, which I shall have occasion to use, to avoid circumlocution: and here it may not be improper to observe, that most of our errors in reasoning have arisen from want of strict attention to this circumstance, the accurate definition of those terms which we use in our reasoning. We may use what terms we please, provided we accurately define them, and adhere strictly to the definition. On this depends the excellence and certainty of the mathematical sciences. The terms are few, and accurately defined; and in their different chains of reasoning mathematicians adhere with the most scrupulous strictness to the original definition of the terms. If the same method were made use of in reasoning on other subjects, they would approach to the mathematics in simplicity and in truth, and the science of medicine in particular would be stripped of the heaps of learned rubbish which now encumber it, and would appear in true and native simplicity. Such is the method I propose to follow: I am certain of the rectitude of the plan; of the success of the reasoning it does not become me to judge.

When the excitability is in such a state as to be very susceptible of the action of external powers, I shall call it abundant or accumulated; but when it is found in a state not very capable of receiving their action, I say it is deficient or exhausted. Let no one however suppose that by these terms I mean to hint in the least at the nature of the excitability. I do not mean by them that it is really at one time increased in quantity or magnitude, and at another time diminished: its abstract nature is by no means attempted to be investigated. These or similar terms the poverty or imperfection of language obliges us to use. We know nothing of the nature of the excitability or vital principle, and by the terms here used I mean only to say, that the excitability is sometimes easily acted on by the external powers, and then I call it abundant or accumulated; at other times the living body is with more difficulty excited, and then I say the vital principle or excitability, whatever it may be, is deficient or exhausted.

On examination we shall find the laws by which external powers act on living bodies to be the following.

First, when the powerful action of the exciting powers ceases for some time, the excitability accumulates, or becomes more capable of receiving their action, and is more powerfully affected by them.

If we examine separately the different exciting powers which act on the body, we shall find abundant confirmation of this law. Besides the exciting powers which act on the body, which I mentioned; viz. heat, food, and air, there are several others, such as light, sound, odorous substances, &c. which will be examined in their proper places. These powers, acting by a certain impulse, and producing a vigorous action of the body, are called stimulants, and life we shall find to be the effect of these and other stimulants acting on the excitability.

The stimulus of light, though its influence in this respect is feeble, when compared with some other external powers, yet has its proportion of force. This stimulus acts upon the body through the medium of the organ of vision. Its influence on the animal spirits strongly demonstrates its connexion with animal life, and hence we find a cheerful and depressed state of mind in many people, and more especially in invalids, to be intimately connected with the presence or absence of the sun. Indeed to be convinced of the effects of light we have only to examine its influence on vegetables. Some of them lose their colour when deprived of it, many of them discover a partiality to it in the direction of their flowers; and all of them perspire oxygen gas only when exposed to it; nay it would seem that organization, sensation, spontaneous motion, and life, exist only at the surface of the earth, and in places exposed to light. Without light nature is lifeless, inanimate, and torpid.

Let us now examine if the action of light upon the body is subject to the law that has been mentioned. If a person be kept in darkness for some time, and then be brought into a room in which there is only an ordinary degree of light, it will be almost too oppressive for him, and will appear excessively bright; and if he have been kept for a considerable time in a very dark place, the sensation will be very painful. In this case, while the retina or optic nerve was deprived of light, its excitability accumulated, or became more easily affected by light: for if a person go out of one room into another, which has an equal degree of light, he will perceive no effect.

You may convince yourselves of the truth of this law, by a very simple experiment; shut your eyes, and cover them for a minute or two with your hand, and endeavour not to think of the light, or what you are doing; then open them, and the daylight will for a short time appear brighter.

If you look attentively at a window for about two minutes, then cast your eyes upon a sheet of white paper, the shape of the window frames will be perfectly visible upon the paper; those parts which express the wood work appearing brighter than the other parts. The parts of the optic nerve on which the image of the frame falls, are covered by the wood work from the action of the light; the excitability of these parts will therefore accumulate; and the parts of the paper which fall upon them must of course appear brighter.

If a person be brought out of a dark room where he has been confined, into a field covered with snow, when the sun shines, it has been known to affect him so much as to deprive him of sight altogether.

This law is well exemplified when we come into a dark room in the day time. At first we can see nothing; but with the absence of light the excitability accumulates, and we begin to have an imperfect glimpse of the objects around us; after a while the excitability of the retina is so far accumulated, and we become so sensible of the feeble light reflected from the surfaces of bodies, that we can discern their shapes, and sometimes even their colours.

Let us next consider what happens with respect to heat, which is a uniform and active stimulus in promoting life. The extensive influence of heat upon animal life is evident from its decay and suspension during winter, in certain animals, and from its revival upon the approach and action of the vernal sun.

If this stimulus is for some time abstracted from the whole body, or from any part, the excitability accumulates, or, in other words, if the body has been for some time exposed to cold, it is more liable to be affected by heat afterwards applied. Of this also you may be convinced by an easy experiment. Put one of your hands into cold water, and then put both into water which is considerably warm: the hand which has been in the cold water will feel much warmer than the other. If you handle some snow in one hand while you keep the other in the bosom, that it may be of the same heat with the body, and then bring both within the same distance of the fire, the heat will affect the cold hand infinitely more than the warm one. This is a circumstance of the utmost importance, and ought always to be carefully attended to. When a person has been exposed to a severe degree of cold for some time, he ought to be cautious how he comes near a fire, for his excitability will be so much accumulated that the heat will act very violently, often producing a great degree of inflammation, and even sometimes of mortification. This is a very common cause of chilblains, and other similar inflammations. When the hands, or any other parts of the body, have been exposed to a violent cold, they ought first to be put in cold water, or even rubbed with snow, and exposed to warmth in the gentlest manner possible.

The same law regulates the action of food, or matters taken into the stomach: if a person have for some time been deprived of food, or have taken it in small quantity, whether it be meat or drink, or if he have taken it of a less stimulating quality, he will find that when he returns to his ordinary mode of life it will have more effect upon him than before he lived abstemiously.

Persons who have been shut up in a coal work, from the falling in of the pit, and have had nothing to eat for two or three days, have been as much intoxicated by a bason of broth, as a person in common circumstances with two or three bottles of wine.

This circumstance was particularly evident among the poor sailors who were in the boat with Captain Bligh after the mutiny. The Captain was sent by government to convey some plants of the bread fruit tree from Otaheite to the West Indies: soon after he left Otaheite the crew mutinied, and put the captain and most of the officers, with some of the men, on board the ship's boat, with a very short allowance of provisions, and particularly of liquors, for they had only six quarts of rum, and six bottles of wine, for nineteen people, who were driven by storms about the south sea, exposed to wet and cold all the time, for nearly a month; each man was allowed only a teaspoonful of rum a day, but this teaspoonful refreshed the poor men, benumbed as they were with cold, and faint with hunger, more than twenty times the quantity would have done those who were warm and well fed; and had it not been for the spirit having such power to act upon men in their condition, they never could have outlived the hardships they experienced. All these facts, and many others which might be brought forward, establish, beyond dispute, the truth of the law I mentioned; viz. that when the powerful action of the exciting powers ceases for some time, the excitability accumulates, or becomes more capable of receiving their actions, and is more powerfully affected by them.

When the legs or arms have for some time been exposed to cold, the slightest exertion, or even the stimulus of a gentle heat, throws the muscles into an inordinate action or cramp. The glow of the skin, in coming out of a cold bath, may be explained on the same principle. The heat of the skin is diminished by the conducting power of the water, in consequence of which the excitability of the cutaneous vessels accumulates; and the same degree of heat afterwards applied, excites these now more irritable vessels to a great degree of action.

On this principle depends the supposed stimulant or tonic powers of cold, the nature of whose action has been much mistaken by physicians and physiologists. Heat is allowed to be a very powerful stimulus; but cold is only a diminution of heat; how then can cold act as a stimulus? In my opinion it never does; but its effects may be explained by the general law which we have been investigating. When a lesser stimulus than usual has been applied to the body, the excitability accumulates, and is then affected by a stimulus even less than that which, before this accumulation, produced no effect whatever. The cold only renders the body more subject to the action of heat afterwards applied, by allowing the excitability to be accumulated. No person, I believe, ever brought on an inflammation, or inflammatory complaint, by exposure to cold, however long might have been that exposure, or however great the cold; but if a person have been out in the cold air, and afterwards come into a warm room, an inflammatory complaint will most probably be the consequence.

Indeed coming out of the cold air into a moderately warm room generally produces a lively and continued warmth in the parts that have been exposed.

The second general law is, that when the exciting powers have acted with violence for a considerable time, the excitability becomes exhausted, or less fit to be acted on; and this we shall be able to prove by a similar induction.

Let us first examine the effects of light upon the eye: when it has acted violently for some time on the optic nerve, it diminishes the excitability of that nerve, and renders it incapable of being affected by a quantity of light, that would at other times affect it. When we have been walking out in the snow, if we come into a room, we shall scarcely be able to see any thing for some minutes.

If you look stedfastly at a candle for a minute or two, you will with difficulty discern the letters of a book which you were before reading distinctly. When our eyes have been exposed to the dazzling blaze of phosphorus in oxygen gas, we can scarcely see any thing for some time afterwards, and if we look at the sun, the excitability of the optic nerve is so overpowered by the strong stimulus of his light, that nothing can be seen distinctly for a considerable time. If we look at the setting sun, or any other luminous object of a small size, so as not greatly to fatigue the eye, this part of the retina becomes less sensible to smaller quantities of light; hence when the eyes are turned on other less luminous parts of the sky, a dark spot is seen resembling the shape of the sun, or other luminous object on which our eyes have been fixed.

On this account it is that we are some time before we can distinguish objects in an obscure room, after coming from broad daylight, as I observed before.

We shall next consider the action of heat. Suppose water to be heated to 90 degrees, if one hand be put into it, it will appear warm; but if the other hand be immersed in water heated to 120 degrees, and then put into the water heated to 90 degrees, that water will appear cold, though it will still feel warm to the other hand: for the excitability of the hand has been exhausted, by the greater stimulus of heat, to such a degree as to be insensible of a less stimulus.

Before we go into a warm bath, the temperature of the air may seem very warm and pleasant to the body, even though exposed naked to it; but after we have remained for some time in the warm bath, we feel the air, when we come out, very cool and chilling, though it is of the same temperature as before; for the hot water exhausts the excitability of the vessels of the skin, and renders them less capable of being affected by a smaller degree of heat. Thus we see that the effects of the hot and cold bath are different and opposite; the one debilitates by stimulating, and the other produces stimulant or tonic effects by debilitating. This seeming paradox may, however, be easily explained by the principles we have laid down; and though the hot and cold bath produce such different effects, yet it is only the same fluid, with a small variation in the degree of temperature; but these effects depend on the temperature of our body being such, that a small decrease of it will produce an accumulation of excitability, while a small increase will exhaust it.

I shall next proceed to examine the effects of the substances taken into the stomach; and as the effects of spirituous and vinous liquors are a little more remarkable than those of food, I shall first begin with them.

A person who is not accustomed to take these liquors, will be intoxicated by a quantity that will produce no effect upon one who has been some time accustomed to take them; and when a person has used himself to these stimulants for some time, the ordinary powers which in common support life, will not have their proper effects upon him, because his excitability has been, in some measure, exhausted by these stimulants.

The same holds good with respect to tobacco and opium; a person accustomed to take opium, or smoke tobacco, will not be affected by a quantity that would completely intoxicate one not used to them, because the excitability has been so far exhausted by the use of those stimulants, that it cannot be acted on by a smaller quantity.

That tobacco or opium act in the same manner as wine or spirits, scarce needs any illustration. In Turkey they intoxicate themselves with opium, in the same way that people in this country do with wine and spirits; and those who have been accustomed to take this drug for a considerable time, feel languid and depressed when they are deprived of it for some time; they repair to the opium houses, as our dram drinkers do to the gin shops in the morning, sullen, dejected, and silent; in an hour or two, however, they are all hilarity. This shows the effects of opium to be stimulant. Tobacco intoxicates those who are not accustomed to it, and in those who are, it produces a serene and composed state of mind by its stimulating effects. Like opium and fermented liquors it exhausts the excitability, and leaves the person dejected, and all his senses blunted, when its stimulant effects are over.

That what is more properly called food acts in the same way as the substances I have just examined, is evident from the fact which I mentioned some time ago, that persons whose excitability has been accumulated, by their being deprived of food for some days, have been intoxicated by a bason of broth.

These facts, with innumerable others which will easily suggest themselves, prove, beyond doubt, the truth of the second law, namely, that when the exciting powers have acted violently, or for a considerable time, the excitability is exhausted, or less fit to be acted on.

Besides the stimulants which I have mentioned, there are several others which act upon the body, many of which will hereafter be considered: but all act according to this law; when their action has been suspended or diminished, the excitability of the organ on which they act becomes accumulated, or more easily affected by their subsequent action; and, on the contrary, when their action has been violent, or long continued, the excitability becomes exhausted, or less fit to receive their actions.

Among the stimulants acting on the body, we may mention sound, which has an extensive influence on human life. I need not mention here its numerous natural, or artificial sources, as that has been fully done in a preceding lecture. The effect of music, in stimulating and producing a state of mind approaching to intoxication, is universally known. Indeed the influence of certain sounds in stimulating, and thereby increasing, the powers of life, cannot be denied. Fear produces debility, which has a tendency to death. Sound obviates this debility, and restores to the system its natural degree of excitement. The schoolboy and the clown invigorate their trembling limbs, by whistling, or singing, as they pass by a country churchyard, and the soldier feels his departing courage recalled in the onset of a battle, by the "spirit stirring drum."

Intoxication is generally attended with a higher degree of life or excitement than is natural. Now sound will produce this effect with a very moderate portion of fermented liquor; hence we find persons much more easily intoxicated and highly excited at public entertainments, where there is music and loud talking, than in private companies, where no auxiliary stimulus is added to that of wine.

Persons who are destitute of hearing and seeing, possess life in a more languid state than other people; which is, in a great degree, owing to the want of the stimulus of light and noise.

Odours have likewise a very sensible effect in promoting animal life. The effects of these will appear obvious in the sudden revival of life, which they produce, in cases of fainting. The smell of a few drops of hartshorn, or even a burnt feather, has frequently, in a few minutes, restored the system from a state of weakness, bordering upon death, to an equable and regular degree of excitement.

All these different stimuli undoubtedly produce the greatest effects upon their proper organs; thus the effect of light is most powerful on the eye; that of sound on the ear; that of food on the stomach, &c. But their effects are not confined to these organs, but extended over the whole body. The excitability exists, one and indivisible, over the whole system; we may call it sensibility, or feeling, to enable us to understand the subject. Every organ, or indeed the whole body, is endowed with this property in a greater or less degree, so that the effects produced by any stimulus, though they are more powerful on the part where they are applied, affect the whole system: odours afford an instance of this; and the prick of a pin in the finger, produces excitement, or a stimulant effect, over the whole body.

From what has been said, it must be evident that life is the effect of a number of external powers, constantly acting on the body, through the medium of that property which we call excitability; that it cannot exist independent of the action of these stimuli; when they are withdrawn, though the excitability does not instantly vanish, there is no life, no motion, but the semblance of death. Life, therefore, is constantly supported by, and depends constantly on, the action of external powers on the excitability; without excitability these stimulants would produce no effect, and whatever may be the nature of the excitability, or however abundant it may be, still, without the action of external powers, no life is produced.

From what has been said, we may see the reason why life is in a languid state in the morning: It acquires vigour by the gradual and successive application of stimuli in the forenoon: It is in its most perfect state about midday, and remains stationary for some hours: From the diminution or exhaustion of the excitability, it lessens in the evening, and becomes more languid at bed time; when, from defect of excitability, the usual exciting powers will no longer produce their effect, a torpid state ensues, which we call sleep, during which, the exciting powers cannot act upon us; and this diminution of their action allows the excitability to accumulate; and, to use the words of Dr. Armstrong,

"Ere morn the tonic, irritable nerves
Feel the fresh impulse, and awake the soul."

LECTURE X. THE LAWS OF ANIMAL LIFE, CONTINUED.

In the last lecture I began to investigate the laws by which living bodies are governed, and the effects produced by the different exciting powers, which support life, upon the excitability, or vital principle. The facts which we examined led us to two conclusions, which, when properly applied, we shall find will explain most of the phenomena of life, both in health, and in disease. The conclusions alluded to, are these: when the exciting powers have acted more feebly, or weakly, than usual, for some time; or when their action is withdrawn, the excitability accumulates, and becomes more powerfully affected by their subsequent action. And, on the contrary, when the action of these powers has been exerted with violence, or for a considerable time, the excitability becomes exhausted, and less fit to receive their actions.

A number of facts were mentioned in proof of these conclusions, and a great number more might have been brought forwards, could it have served any other purpose than to have taken up our time, which I hope may be better employed.

This exhaustion of the excitability, by stimulants, may either be final, or temporary. We see animals, while the exciting powers continue to act, at first appear in their greatest vigour, then gradually decay, and at last come into that state, in which, from the long continued action of the exciting powers, the excitability is entirely exhausted, and death takes place.

We likewise see vegetables in the spring, while the exciting powers have acted on them moderately, and for a short time, arrayed in their verdant robes, and adorned with flowers of many mingling hues; but as the exciting powers, which support their life, continue to be applied, and some of them, for instance heat, as the summer advances, become increased, they first lose their verdure, then grow brown, and at the end of summer cease to live: because their excitability is exhausted by the long continued action of the exciting powers: and this does not happen merely in consequence of the heat of the summer decreasing, for they grow brown, and die, even in a greater degree of heat than that which in spring made them grow luxuriantly. In some of the finest days of autumn, in which the sun acts with more power than in the spring, the vegetable tribe droop, in consequence of this exhausted state of their excitability, which renders them nearly insensible of the action, even of a powerful stimulus.

These are examples of the final or irreparable exhaustion of the excitability; but we find also that it may be exhausted for a time, and accumulated again. Though the eye has been so dazzled by the splendour of light, that it cannot see an object moderately illuminated, yet if it be shut for some time, the excitability of the optic nerve will accumulate again, and we shall again be capable of seeing with an ordinary light.

We find also that we are not always equally capable of performing the functions of life. When we have been engaged in any exertion, either mental or corporeal, for some hours only, we find ourselves languid and fatigued, and unfit to pursue our labours much longer.

If in this state several of the exciting powers are withdrawn, particularly light and noise, and if we are laid in a posture which does not require much muscular exertion, we soon fall into that state which nature intended for the accumulation of the excitability, and which we call sleep. In this state many of the exciting powers cannot act upon us, unless applied with some violence, for we are insensible to their moderate action. A moderate degree of light, or a moderate noise, does not affect us, and the power of thinking, which very much exhausts the excitability, is in a great measure suspended. When the action of these powers has been suspended for six or eight hours, the excitability is again capable of being acted on, and we rise fresh and vigorous, and fit to engage in our occupations.

Sleep then is the method which nature has provided to repair the exhausted constitution, and restore the vital energy. Without its refreshing aid, our worn out habits would scarcely be able to drag on a few days, or at most, a few weeks, before the vital spring would be quite run down: how properly therefore has our great poet called sleep "the chief nourisher in life's feast!"

From the internal sensations, often excited, it is natural to conclude, that the nerves of sense are not torpid during sleep, but that they are only precluded from the perception of external objects, by the external organs being in some way or other rendered unfit to transmit to them the impulses of bodies during the suspension of the power of volition; thus the eyelids are closed, in sleep, to prevent the impulse of light from acting on the optic nerve; and it is very probable that the drum of the ear is not stretched; it seems likewise reasonable to conclude, that something similar happens to the external apparatus of all our organs of sense, which may make them unfit for their office of perception during sleep.

The more violently the exciting powers have acted, the sooner is sleep brought on, because the excitability is sooner exhausted, and therefore sooner requires the means of renewing it: and, on the contrary, the more weakly these powers have acted, the less are we inclined to sleep. Instances of the first are, excess of exercise, strong liquors, or study; and of the latter, an under or deficient proportion of these.

A person who has been daily accustomed to much exercise, whether mental or corporeal, if he omit it, will find little or no inclination to sleep; this state may however be induced by taking some diffusible stimulus, as a little spirits and water, or opium, which seem to act entirely by exhausting the excitability, to that degree which is compatible with sleep, and, when the stimulant effect of these substances are over, the person soon falls into that state.

But though the excitability may have been sufficiently exhausted, and the action of external powers considerably moderated, yet there are some things within ourselves, which often stimulate violently, and prevent sleep, such as pain, thirst, and strong passions and emotions of the mind. These all tend to drive away sleep, by their vehement stimulating effect, which still has power to rouse the excitability to action, though it has been considerably exhausted. The best method of inducing sleep, in these cases, is to endeavour to withdraw the mind from these impressions, particularly from uneasy emotions, by employing it on something that makes a less impression, and which does not require much exertion, or produce too much commotion; such as counting to a thousand, or counting drops of water which fall slowly; by listening to the humming of bees, or the murmuring of a rivulet. Virgil describes a situation fitted to induce sleep, most beautifully, in the following words.

> "Fortunate senex, hic inter flumina nota,
> Et fontes sacros, frigus captabis opacum.
> Hinc tibi, quae semper vicino ab limite sepes
> Hyblaeis apibus florem depasta salicti,
> Seape levi somnum suadebit inire susurro."

In infancy much sleep is required; the excitability, being then extremely abundant, is soon exhausted by external stimulants, and therefore soon requires renewing or accumulating; on this account, during the first five or six months of their life, children require this mode of renewing their exhausted excitability several times in the day; as they advance in years, and as this excess of excitability is exhausted by the application of stimulants, less sleep is required: in the prime of life least of all is necessary. There is great difference however, in this respect, in different constitutions. Some persons are sufficiently refreshed by three or four hours sleep, while others require eight or ten hours. More however depends, in my opinion, on the mode of living. Those who indulge in the use of spirituous or fermented liquors, which exhaust the excitability to a great degree, require much more sleep than those who are content with the crystal stream. The latter never feel themselves stupid or heavy after dinner, but are immediately fit to engage in study or business. As age advances, more sleep is again required; and the excitability at last becomes so far exhausted, and the system so torpid, that the greatest portion of gradually expiring life is spent in sleep.

Temperance and exercise are the most conducive to sound healthy sleep, hence the peasant is rewarded, for his toil and frugal mode of life, with a blessing, which is seldom enjoyed by those whom wealth renders indolent and luxurious. The poor in the country enjoy sound and sweet sleep: forced by necessity to labour, their excitability becomes exhausted in a proper and natural manner, and they retire to rest early in the evening. Their sleep is generally sound, and early in the morning they find themselves recruited, and in a state fit to resume their daily labour. The blooming complexion, strength, and activity, of these hardy children of labour, who recruit their wearied limbs on pallets of straw, form a striking contrast with the pallid and sickly visage, and debilitated constitution of the luxurious and wealthy, who convert night into day, and court repose in vain on beds of down. Nature undoubtedly intended that we should be awake, and follow our occupations, whether of pleasure or business, during the cheering light of day, and take repose when the sun withdraws his rays. All other animals, and even vegetables, obey the command of nature: man alone is refractory; but nature's laws are never violated with impunity. Dr. Mackenzie very properly observes, that those who sleep long in the morning, and sit up all the night, injure the constitution without gaining time: and those who do this merely in compliance with fashion, ought not to repine at a fashionable state of bad health.

From what has been said, it is evident that, in order to enjoy sound sleep, our chambers should be free from noise, dark, and moderately cold; because the stimulant effects of noise, light, and heat, prevent the accumulation of excitability: and as we shall afterwards see that this accumulation depends on free respiration, and the introduction of oxygen by that means into the system, our bed rooms ought to be large and airy, and, in general, the beds should not be surrounded by curtains. We may from this likewise see the reason why it is so desirable to sleep in the country, even though we are obliged to spend the day in town.

These observations on sleep have however led me a little from the direct road; but I thought they could not be better introduced than here. I shall now return to the subject of our more immediate inquiry.

By induction we have discovered two of the principal laws by which living bodies are governed: the first is, that when the ordinary powers which support life have been suspended, or their action has been lessened for a time, the excitability, or vital principle, accumulates, or becomes more fit to receive their actions; and secondly, when these powers have acted violently, or for a considerable time, the excitability is exhausted, or becomes less fit to receive their actions. There are therefore three states in which living bodies exist. First, a state of accumulated excitability. Secondly, a state of exhausted excitability. Thirdly, when the excitability is in such a state as to produce the strongest and most healthy actions, when acted upon by the external powers.

From what has been said, it must be evident that life depends continually on the action of external powers on the excitability, and that by their continued action, if they be properly regulated, the excitability will be gradually, and insensibly exhausted, and life will be resigned into the hands of him who gave it, without a struggle, and without a groan.

We see then that nature operates in supporting the living part of the creation, by laws as simple and beautiful as those by which the animated world is governed. In the latter we see the order and harmony which is observed by the planets, and their satellites, in their revolution round the great source of heat and light;

> "— — — — — all combin'd And ruled unerring, by that Single Power, Which draws the stone projected, to the ground.

In the animated part of the creation, we observe those beautiful phenomena which are exhibited by an almost infinite variety of individuals; all depending upon, and produced by one simple law; the acting of external powers upon their excitability.

I cannot express my admiration of the wisdom of the Creator better than in the words of Thomson.

"O unprofuse magnificence divine!
O wisdom truly perfect! thus to call
From a few causes such a scheme of life;
Effects so various, beautiful, and great."

Life then, or those functions which we call living, are the effects of certain exciting powers acting on the excitability, or property distinguishing living from dead matter. When these effects, viz. the functions, flow easily, pleasantly, and completely, from the action of these powers, they indicate that state which we call health.

We may therefore, as we before hinted, distinguish three states of the irritable fibre, or three different degrees of excitability, of which the living body is susceptible.

1. The state of health which is peculiar to each individual, and which has been called by Haller, and other physiologists, the tone of the fibre. This is produced by a middle degree of stimulus acting upon a middle degree of excitability: and the effect produced by this action, we call excitement.

2. The state of accumulation, produced by the absence or diminished action of the accustomed stimuli.

3. The state of exhaustion, produced by the too powerful action of stimuli; and this may be produced either by the too powerful, or long continued action of the common stimulants which support life, such as food, air, heat, and exercise; or it may be caused by an application of stimulants, which act more powerfully on the excitability, and which exhaust it more quickly, such as wine, spirits, and opium, musk, camphor, and various other articles used in medicines.

The state of health, or tone, if we use that term, consists therefore in a certain quantity or energy of excitability necessary to its preservation. To maintain this state, the action of the stimuli should be strong enough to carry off from the body the surplus of this irritable principle. To obtain this end, a certain equilibrium is necessary between the excitability and the stimuli applied, or the sum of all the stimuli acting upon it must be always nearly equal, and sufficient to prevent an excess of excitability, but not so strong as to carry off more than this excess. It is in this equilibrium between the acting stimuli and the excitability, that the health, or tone of the living body consists.

When the sum of the stimuli, acting on the body, is so small, as not to carry off the excess of excitability, it accumulates, and diseases of irritability are produced. Of this nature are those diseases to which the poor are often subject, and which will be particularly considered hereafter.

When the sum of the stimuli acting on the body, is too great, it is deprived not only of the excess of excitability, but also of some portion of the irritable principle necessary for the tone of the body: or, to speak more distinctly, the body loses more excitability than it receives, and of course must, in a short time, be in a state of exhaustion. This gives rise to diseases which afflict drinkers, or those who indulge in any kind of intemperance, or persons born in climates where the temperature is moderate, but who emigrate to those which are much warmer.

Thus we have endeavoured, after the example of Dr. Brown, to ascertain the cause of the healthy state, before the causes of diseases were investigated; and though this is contrary to the general practice, yet it must be evident to every one, that unless we are acquainted with the causes of good health, it will be impossible for us to form any estimate of those variations from that state, called diseases: hence it is that a number of diseases, which have been brought on merely by the undue action of the exciting powers, such as gout, rheumatism, and the numerous trains of nervous complaints, which were by no means understood, may be easily and satisfactorily explained, and as easily cured, by restoring the proper action of these powers, and bringing the excitability to its proper state. As this theory, therefore, is so important, not only in respect to the preservation of health, which nearly concerns every individual, but to the cure of diseases, which is the province of the physician, I have endeavoured to explain it as fully and minutely as possible; to make it still plainer we may perhaps make use of the following illustration.

Suppose a fire to be made in a grate or furnace, filled with a kind of fuel not very combustible, and which could only be kept burning by means of a machine, containing several tubes placed before it, and constantly pouring streams of air into it. Suppose also a pipe to be fixed in the back of the chimney, through which a constant supply of fresh fuel is gradually let down into the grate, to repair the waste occasioned by the combustion kept up by the air machine.

The grate will represent the human body; the fuel in it the life or excitability, and the tube behind, supplying fresh fuel, will denote the power of all living systems, constantly to regenerate or produce excitability; the air machine, consisting of several tubes, may denote the various stimuli applied to the excitability of the body; the flame produced in consequence of that application, represents life; the product of the exciting powers acting upon the excitability.

Here we see, that flame, like life, is drawn forth from fuel by the constant application of streams of air, poured into it from the different tubes of the machine. When the quantity of air poured in through these different tubes is sufficient to consume the fuel as it is supplied, a constant and regular flame will be produced: but if we suppose that some of them are stopped, or that they do not supply a sufficient quantity of air, then the fuel will accumulate, and the flame will be languid and smothered, but liable to break out with violence, when the usual quantity of air is supplied.

On the contrary, if we suppose a greater quantity of air to rush through the tubes, then the fuel will be consumed or exhausted faster than it is supplied; and in order therefore to reduce the combustion to the proper degree, the quantity of air supplied must be diminished, and the quantity of fuel increased.

If we suppose one of the tubes, instead of common air, to supply oxygen gas, it will represent the action of wine, spirits, ether, opium, and other powerful stimulants upon the body: a bright and vivid flame will be produced, which however will only be of short duration, for the fuel will be consumed faster than it is supplied, and a state of exhaustion will take place.

We may carry this illustration still further, and suppose that the air tubes exhaust the fuel every day faster than it can be supplied, then it will be necessary at night to stop up some of the tubes, so that the expense of fuel may be less than its supply, in order to make up for the deficiency. When this is made up, the tubes may in the morning be opened, and the combustion carried on during the day as usual. This will illustrate the nature of sleep. In speaking of this subject, it was observed, that the more violently the exciting powers have acted, the sooner is sleep brought on; because the excitability is sooner exhausted. In the same way the more the air rushes through the tubes, the sooner will the fuel be consumed, and want replenishing. When the exciting powers have acted feebly, a person feels no inclination to sleep, because the excitability is not exhausted to the proper degree, and therefore does not want accumulating. But any diffusible stimulus, as spirits, or opium, will soon exhaust it to the proper degree.

In the same way, if the air have not passed rapidly through the tubes, the fuel will not be exhausted: but it may be brought to a proper degree of exhaustion by the application of oxygen gas.

When the air which nourishes the flame is so regulated, that it consumes the fuel as it is supplied, but no faster, a clear and steady flame will be kept up, which will go on as long as the fuel lasts, or the grate resists the action of the fire: but at last when the fuel, which we do not suppose inexhaustible, is burnt out, the fire must cease.

In the same manner, if the different exciting powers which support life were properly regulated, all the functions of the body would be properly performed, and we should pass our life in a state of health, seldom known to any but savages, and brute animals not under the dominion of man, who regulate these powers merely by the necessities of nature.

When air is applied in too great quantity, and especially if some of the tubes convey oxygen gas, then a violent combustion and flame is excited, which will, in all probability, consume or burn out the furnace or grate, or if it do not, it will burn out the fuel, and thus exhaust itself.

In like manner, if the stimulants which support life be made to act too powerfully, and particularly if any powerful stimulus, not natural to the body, such as wine or spirits, be taken in too great quantity, a violent inflammatory action will be the consequence, which may destroy the human machine: but if it do not, it will exhaust the excitability, and thus bring on great debility.

This analogy might be pursued further, but my intention was solely to illustrate some of the outlines of our theory, by a comparison which may facilitate the conception of the manner in which external powers act on living bodies. The different powers which support life, and without whose action we are unable to exist, such as heat, food, air, &c. have been very improperly called nonnaturals, a term which is much more applicable to those substances which we are daily in the habit of receiving into the system, which excite it to undue actions, and which nature never intended we should receive; such as spirituous and fermented liquors, and high seasoned foods. In the preceding illustration, I have spoken of a tube, as constantly pouring in fresh fuel, because it was not easy otherwise to convey a familiar idea of the power which all living systems possess of renewing their excitability, when exhausted. The excitability is an unknown somewhat, subject to peculiar laws, some of which we have examined, but whose different states we are obliged to describe, though, perhaps, inaccurately, by terms borrowed from the qualities of material substances.

Though Dr. Brown very properly declined entering into the consideration of the nature of excitability, or the manner in which it is produced, the discoveries which have been made in chemistry since his time, have thrown great light on the subject, and it is now rendered highly probable that the excitability or vital principle, is communicated to the body by the circulation, and is intimately connected with the process of oxidation.

Many circumstances would tend to show, that a strict connexion exists between the reception of oxygen into the body, and the vital principle.

When an animal has been killed by depriving it of oxygen gas, the heart and other muscles, and indeed the whole system, will be found completely to have lost its excitability. This is not the case when an animal is killed in a different manner. When an animal is shot, or killed in the common manner, by bleeding to death, if the heart be taken out, it will contract for some hours, on the application of stimulants. But this is not the case with an animal that has been drowned, or killed by immersion in carbonic acid, azotic, or hydrogenous gases; in these last instances, the heart either does not contract at all, or very feebly, on the application of the strongest stimulants.

We have already seen that oxygen unites with the blood in the lungs, during respiration: by the circulation of the blood it is distributed to every part of the system, and we shall find, that in proportion to its abundance is the excitability of the body. In proof of this, I shall relate some facts and experiments.

Dr. Girtanner injected a quantity of very pure oxygen gas into the jugular vein of a dog: the animal raised terrible outcries, breathed very quickly, and with great difficulty: by little and little his limbs became hard and stiff, he fell asleep, and died in the course of a few minutes afterwards. It ought here to be observed, that any of the gases, or almost any fluid, however mild, when thus suddenly introduced into the circulating system, generally, and speedily, occasions death.

On opening the chest, the heart was found more irritable than ordinary, and its external contractions and dilatations continued for more than an hour: the right auricle of the heart, which usually contains black venous blood, contained, as well as the right ventricle, a quantity of blood of a bright vermilion colour; and all the muscles of the body were found to be more than usually irritable. This experiment not only proves that the vermilion colour of the blood proceeds from oxygen, but likewise seems to show, that oxygen is the cause of excitability.

A quantity of azotic gas, which had been exposed for some time to the contact of lime water, in order to separate any carbonic acid gas it might contain, was injected into the jugular vein of a dog. The animal died in twenty seconds. Upon opening the chest, the heart was found filled with black and coagulated blood: this organ, and most of the muscles had nearly lost the whole of their irritability, for they contracted but very weakly, on the application of the strongest stimulants.

A quantity of carbonic acid gas was injected into the jugular vein of a dog: the animal became sleepy, and died in about a quarter of an hour: the heart was found filled with black and coagulated blood, and had lost the whole of its irritability; neither it, nor any of the muscles producing any contractions, upon the application of stimulants.

Humboldt likewise mentions a curious fact, which tends strongly to confirm this idea. When the excitability of the limb of a frog had been so far exhausted, by the application of zinc and silver, that it would produce no more contractions, on moistening it with oxygenated muriatic acid, the contractions were renewed.

After the excitability of the sensitive plant (mimosa pudica) had been so far exhausted, by irritation, that it ceased to contract, when further irritated, I restored this excitability, and brought it to a very high degree of irritability, by moistening the earth in which it grew with oxygenated muriatic acid. Seeds likewise vegetate more quickly when moistened with this acid, than when they are not.

In short, we shall find, first, that every thing which increases the quantity of oxygen in organized bodies, increases at the same time their excitability.

Secondly, That whatever diminishes the quantity of oxygen, diminishes the excitability.

The excitability of animals, made to breathe oxygen gas, or to take the oxygenated muriate of potash, or acid fruits, is very much increased.

On the contrary, when persons have inspired carbonic acid, or azotic gas, or have taken into the system substances which have a strong affinity for oxygen, and therefore tend to abstract it, such as hydrogen, and spirits, the excitability becomes very much diminished.

When we sleep, in consequence of the excitability being exhausted, the breathing becomes free, and a great quantity of oxygen is received by the lungs, and combined with the blood, while very little of it becomes exhausted by the actions of the body, for none, excepting those which are called involuntary motions, are carried on during sound sleep: so that in a few hours the body recovers the excitability which it had lost: it is again sensible of the impressions of external objects, and with the return of light we wake.

These facts afford satisfactory proofs that the excitability of the body is proportioned to the oxygen which it receives: but in what manner it produces this state of susceptibility, and how it is exhausted by stimulants, we have yet to learn.

The following theory may perhaps throw some light upon the subject. I propose it, however, merely as an hypothesis, for we have no direct proofs of it, but it seems to account for many phenomena.

It is now well known, that while the limb of an animal possesses excitability, the smallest quantity of electricity sent along the principal nerve leading to it, produces contractions similar to those produced by the will. This is instanced in the common galvanic experiment with the limb of a frog, which I had formerly occasion to show.

From the effects produced, when a stream of electricity is sent through water, I think it not improbable that hydrogen and electricity may be identical. When a piece of zinc and silver are connected together, and the zinc is put in a situation to decompose water, and oxidate, a current of hydrogen gas will separate from the silver wire, provided this be immersed under water; but when it is not, a current of electricity passes, which is sensible to the electrometer.

Now there appears no greater improbability in the supposition that hydrogen, in a certain state, may be capable of passing through metals, and animal substances, in the form of electricity, and that when it comes in contact with water, which is not so good a conductor, it may combine with caloric, and form hydrogen gas, in which state it becomes incapable of passing through the conductors of electricity: I say there appears no greater improbability in this, than that caloric should sometimes be in such a state, that it will pass through metals, and animal substances, which conduct it, and at other times, as when combined with oxygen or hydrogen, it should form gases, and be then incapable of passing through these conductors of heat. Galvanic effects may be produced by the oxidation of fresh muscular fibre without the aid of metals, and contractions have been thus produced in the limb of an animal; and we have already noticed, that when this contraction ceases, it may be restored, by moistening the limb with oxygenated muriatic acid.

The excitability of the body may, most probably, be conveyed by respiration, and the circulation of the blood, which tend continually to oxidate the different parts: and hydrogen or electricity may be secreted by the brain, and sent along the nerves, which are such good conductors of it, and by uniting with the oxygen of the muscle, may cause it to contract; but as the oxygen will, by this union, be diminished, if the contractions be often repeated, the excitability will thus be expended faster than it can be supplied by the circulation, and will become exhausted. But will facts bear us out in this explanation? To see this, we must examine the chemical nature of the substances which produce the greatest action, and the greatest exhaustion of the vital principle: namely, those which produce intoxication.

Fermented liquors differ from water, in containing carbon and more hydrogen: these produce intoxication: but pure spirits, which contain still more hydrogen, produce a still higher degree of intoxication, and consequent exhaustion of the excitability. Ether, which appears to be little more than condensed hydrogen, probably kept in a liquid state by union with a small quantity of carbon, and which easily expands by caloric into a gas, which very much resembles hydrogen gas, produces a still greater degree of intoxication: so that we see the action produced by different substances, as well as the exhaustion of excitability which follows, is proportioned to the quantity of hydrogen they contain.

There is another circumstance which seems to strengthen this idea. The intoxicating powers of spirits are diminished by the addition of vegetable acids, or substances which contain oxygen, which will counteract the effects of the hydrogen. Thus it is known that the same quantity of spirit, made into punch, will not produce either the same ebriety, or the same subsequent exhaustion, as when simply mixed with water.

Recollect however that I propose this only as a hypothesis: its truth may be confirmed by future observations and experiments, or it may be refuted by them: but it is certainly capable of explaining many of the phenomena, which is one of the conditions required by Newton's first rule of philosophizing.

Heat, and light, and other stimuli, may perhaps exhaust the excitability, by facilitating the combination of oxygen in the fibres with the hydrogen and carbon in the blood.

There are several substances which cause a diminution or exhaustion of the excitability, without producing any previous increased excitement. These substances have by physicians been called sedatives: and though the existence of such bodies is denied by Dr. Brown, yet we are constrained to admit them; nor do their effects seem incapable of being explained on the principle laid down, especially if we call in the aid of chemistry.

Any substance which is capable of combining rapidly with oxygen, and diminishing its quantity, will be a sedative. But the action of some of the animal and vegetable poisons is difficult to explain in the present state of our knowledge; such very minute portions of these produce great exhaustion of the excitability, and even death, that we can scarcely explain their action on the supposition that they combine with the oxygen. They may perhaps act as ferments, and occasion throughout the whole system a new and rapid combination of oxygen with the hydrogenous, carbonic, and perhaps azotic parts of the blood and fluids, and even of the solids, which will speedily destroy the excitability, and even the organization.

Many of the vegetable narcotics, though they will destroy life when given in considerable doses, yet when exhibited in less quantities become very powerful remedies, particularly in cases where the excitability is accumulated, in consequence of which violent spasms and inordinate actions take place, which are very quickly calmed by opium, camphor, musk, asafoetida, ether, &c. medicines that occasion a speedy exhaustion of the excitability. In diseases of exhaustion, however, these remedies are improper. The indication here is to accumulate the irritability, by the introduction of oxygen, and by the diminution of the action of the stimulants which support life. In this idea too I dissent from Dr. Brown, who taught that diseases of exhaustion are to be cured by stimulants, a little less powerful than those which produced the disease. This subject will however be more fully discussed hereafter.

This doctrine of animal life, which I have been attempting to illustrate, and render familiar, exhibits a new view of the manner in which it is constantly supported. It discovers to us the true means of promoting health and longevity, by proportioning the number and force of stimuli to the age, climate, situation, habits, and temperament, of the human body. It leads us to a knowledge of the causes of diseases: these we shall find consist either in an excessive or preternatural excitement in the whole or part of the human body, accompanied generally with irregular motions, and induced by natural or artificial stimuli, or in a diminished excitement or debility in the whole, or in part. It likewise teaches us that the natural and only efficacious cure of these diseases depends on the abstraction of stimuli, from the whole, or from a part of the body, when the excitement is in excess: and in the increase of their number and force when the contrary takes place.

The light which the discoveries of Galvani, and others who have followed his steps, begin to throw on physiology, promises, when aided by the principles of chemistry, and the knowledge of the laws of life, to produce all the advantages that would result from a perfect knowledge of the animal functions.

From what has been said, it does not seem improbable that muscular contraction may depend upon the combination of oxygen with hydrogen and azote, in consequence of a sort of explosion or discharge produced by nervous electricity. According to this hypothesis, animal motion, at least that of animals analogous to man, would be produced by a beautiful pneumatic structure. This hypothesis, though not perhaps at this moment capable of strict demonstration, seems extremely probable, it being countenanced by every observation and experiment yet made on the subject. It accounts likewise for the perpetual necessity of inhaling oxygen, and enables us to trace the changes which this substance undergoes, from the moment it is received into the system, till the moment it is expelled. By the lungs it is imparted to the blood; by the blood to the muscular fibres; in these, during their contraction, it combines with the hydrogen, and perhaps carbon and azote, to form water and various salts, which are taken up by the absorbents, and afterwards exhaled or excreted. We know the necessity of oxygen to muscular motion, and likewise that this motion languishes when there is a deficiency of the principle, as in sea scurvy. Thus a boundless region of discovery seems to be opening to our view: the science of philosophy, which began with remote objects, now promises to unfold to us the more difficult and more interesting knowledge of ourselves. Should this kind of knowledge ever become a part of general education, then the causes of many diseases being known, and the manner in which the external powers, with which we are surrounded, act upon us, a great improvement not only in health, but in morality must be the consequence.

With respect to its influence on the science of medicine, we may observe that, from the time of Hippocrates till almost the present day, medicine has not deserved the name of a science but, as he called it, of a conjectural art. At present however, by the application of the laws of life, and of the new chemistry, there is beginning to appear in physiology and pathology, something like the simplicity and certainty of truth. In proportion as the laws of animal nature come to be ascertained, the study of them will excite more general attention, and will ultimately prove the most popular, as well as the most curious and interesting branch of philosophy.

This must be productive of beneficial consequences to society, since these truths, once impressed upon the mind by conviction, will operate as moral motives, by which the sum of disease and human misery cannot fail to be greatly diminished.

LECTURE XI. OF THE NATURE AND CAUSES OF DISEASES.

In the two last lectures I have attempted to investigate the laws of life. I now proceed to the most important part of our course, and for which all the preceding lectures were intended to prepare us; I mean the application of the laws of life to explain the nature and causes of diseases, and the methods of curing them, which must always be imperfect, and conjectural, unless the nature of the diseases themselves be well understood.

We have already seen that life is constantly supported by the action of the external powers which surround us, and that if the action of these powers be properly regulated, and at the same time no other powers be suffered to act on the body, we shall enjoy perfect health, but if, on the contrary, the exciting powers which support life, act either too feebly or too powerfully, then the functions will not be performed with precision and vigour, but irregularly; the mind and body will become deranged, and death will often take place many years before the natural period at which that event might be expected.

As health is the greatest blessing which man can enjoy, it is natural to think, that in the early ages of society, when men began to lose sight of the dictates of nature, and feel the torture of disease, they would regard with gratitude those who had contributed towards their relief, and that they would place their physicians among their heroes and their gods. In the early ages, however, diseases would be very few, for it would not be till civilisation had made considerable progress, that such unnatural modes of life as conduce to their production, would take place.

As the first professors of physic knew nothing of the animal economy, and little of the theory of diseases, it is evident that whatever they did, must have been in consequence of mere random trials. Indeed it is impossible that this or any other art could originate in any other manner. Accordingly history informs us that the ancient nations used to expose their sick in temples, and by the sides of highways, that they might receive the advice of every one that passed.

It would take up too much time to pursue the history of medicine from this rude origin, through all its changes and revolutions, till the present time: let it therefore suffice to say, that after various theories had been invented and overturned, and after one age had destroyed the labours of another, though different branches of the healing art, and particularly anatomy, had been enriched with valuable discoveries, still a rational theory was wanting; there was nothing to guide the practitioner in his way, and we may truly say that till the laws of life, which I have been endeavouring to illustrate, were investigated by Dr. Brown, medicine could boast of no theory which had a title to be called philosophical.

The theories of Stahl, Boerhaave, and Cullen, have passed away, and are almost forgotten, but this, which is founded on nature, and on fact, will, like the Newtonian philosophy, last for ever. It has already influenced the practice of medicine, and is taught in almost all the schools of Europe and America. In this country it seems to have had less attention paid to it than it deserved, because its influence was counteracted by the arrogance and profligacy of its author, as if the grossness of a man's manner affected the conclusiveness of his arguments; but this influence did not extend beyond Britain, while the light of his theory illuminated the opposite hemisphere. And when the manner in which he was persecuted is recollected, the liberal mind will allow something to the deep consciousness of neglected merit.

A circumstance much in favour of this doctrine is, that those who understand its principles thoroughly, are guided by it in their practice with a certainty and success before unknown. I say those who understand its principles, for these were not perfectly understood even by the author himself. He first saw with his mind's eye the grand outline of the system, from which, for want of proper reflection, he often drew wrong deductions, and which he often applied improperly. But whatever errors Brown may have committed in the application of his system, and however short his doctrines may fall of a perfect system of medicine, we may venture to predict that the grand outlines will remain unshaken.

From what has been already shown, it must be evident that if the just degree of excitement could be kept up, mankind would enjoy continual health. But it is difficult, if not impossible, to regulate the action of the exciting powers in this equable manner, and if their action is increased, the first effect they produce on the functions is to increase them, and the next is, to render them disturbed or uneasy; or, in other words, to bring on diseases of increased action, or what have been called inflammatory or phlogistic, both of which terms are improper, as they convey false ideas, and are connected with erroneous theories: Dr. Brown has given the name of sthenic to these diseases, from their consisting in increased strength or action, and this is certainly a more appropriate term. On the contrary, when the action of the exciting powers is diminished more than is natural, the functions become languid and disturbed, and by a still further decrease of the action of these powers, they become irregular and inordinate. This state of the body, which is opposite to the former, Dr. Brown has denominated asthenic.

But the stimulant powers may act so powerfully, and exhaust the excitability to such a degree, that they may overstep the bounds of sthenic or inflammatory disease and bring on debility. Debility may therefore arise either from the stimuli acting too weakly, or from a deficient excitability, while the stimulus is not deficient. Debility produced in the former manner is called direct debility, and in the latter indirect debility.

To explain this more clearly, let us take a common instance. If a person by any means be deprived of the proper quantity of food, he will feel himself enfeebled, and the functions will gradually grow more and more languid, and at last become irregular, and be performed with pain. This state is called direct debility. Here is excitability enough, and even too much, for it has accumulated by the subtraction of a stimulus; but here is a deficiency of excitement from defect of stimulus.

If now we suppose that a person, in good health, begins to take a greater quantity of food than usual, and adds a quantity of wine, all the functions will at first be increased in vigour, but at last they will be irregularly performed, and inflammation, with other symptoms of too great excitement, will be the consequence. This state is called sthenic diathesis or disease. But if the stimulant power be pushed still further, the excitability will become gradually exhausted, till at last there will be too little to produce the healthy actions, even though there may be plenty of stimulus. This state of asthenic diathesis is called indirect debility, because it is not produced by directly subtracting the powers which support life, but indirectly, by over stimulating. An instance of this latter state is afforded by that debility which is the consequence of intoxication.

There is a state however between perfect health and disease, which is called predisposition; and in which, though the functions are undisturbed, the slightest cause will bring on disease. Strictly speaking, there is perhaps only one point, or one degree of excitement, at which the health is perfect: the first alterations from this point, on either side, are scarcely perceptible, but if the morbid causes be continued, the functions will become gradually more and more disturbed, till at last they become so uneasy or painful that they are termed disease.

In order to render what has been said still more plain, it may be proper to make use of an illustration by means of numbers: we must recollect however that it is merely for the sake of illustration, for we have not data to enable us to reduce either the excitability, or excitement, or stimulus, to numerical calculation; if we could do this, the science of medicine would be perfect, and we could cure diseases as easily as we could perform any chemical or philosophical experiment. A very principal object however is to understand the nature of predisposition, and the kind of diathesis, whether sthenic or asthenic, to which it inclines: this not only throws light on the nature of the disease, but affords us the only means of preventing it. When a slight uneasiness or predisposition is felt, it is almost impossible to say from our feelings whether it leads to a sthenic or an asthenic state: here we must be guided chiefly by the exciting powers. If we find that these have acted too powerfully; that is, if we have lived freely, been exposed to heat, and perhaps indulged in some of the unnatural stimuli, such as wine and spirits; and particularly if we previously to the present time perceived the functions to go on with more vigour, our spirits and strength greater, before we experienced the slight disturbance of which we complain, we are verging towards sthenic or inflammatory disease, and therefore to prevent the disease we ought immediately to diminish the action of the exciting powers; the quantity of food ought to be diminished, wine and other liquors abstained from, heat carefully avoided; and even the quantity of blood in the circulating system diminished, if the habit is full and the pulse strong.

On the contrary, if the exciting powers have acted more feebly than is natural; that is, if we have lived on a less nourishing diet, or have taken it in less quantity; if we have been long exposed to cold, without alternating with heat, and other debilitating causes; and if at the same time we find the vigour of the functions diminished, though they are not yet become much disturbed, we are verging towards asthenic disease. To prevent which, we must take a more nutritious diet, and join a portion of wine, and perhaps take some tonic medicines. This however ought to be done gradually, for fear of exhausting the excitability, which in these cases is morbidly accumulated.

It must be evident that the great difficulty here is to determine the nature of the predisposition; for if we make a mistake, instead of preventing, we shall accelerate the disease. For instance, the first slight disturbance of the functions which rises from a sthenic state, often resembles those verging towards a state of debility or asthenia. I have seen various instances arising from plethora, or a sthenic state, where the patient complained of depression of spirits, and inability to move; and, in short, from his own account was labouring under asthenic diathesis: but by inquiring carefully into the action of the exciting causes, examining minutely the state of the pulse and of the functions, I have been convinced that the depression of spirits which he felt, and other symptoms of weakness, depended on fullness, and they have been quickly removed by lowering the diet, administering a laxative, or taking a little blood: whereas if, apprehending from the symptoms that he had laboured under debility, I had ordered him a more generous diet and tonic remedies, an inflammatory disease would have been the consequence, which might have terminated in death.

I have seen various instances where patients have complained of this unusual depression, and inability to move: they have shown me prescriptions in which the stimulant or tonic plan was recommended, but instead of any alleviation the symptoms had become worse from their use. This hint was generally sufficient, for if the disease of predisposition had been asthenic, cordials and tonics ought to have relieved it: if, on inquiry, I found the exciting powers had acted too powerfully, I then, without hesitation, had recourse to the debilitating plan, and with the greatest certainty of success. Before I viewed diseases and their causes in this way, I must confess that I often felt great hesitation in practice; and judging merely from symptoms, which are frequently very fallacious, the operation of a remedy often disappointed me, and I could not pretend to predict the event with the certainty that I now can. This observation is of the greatest consequence in the cure both of predisposition and of disease. Though excitement regulates all the phenomena of life, yet the symptoms of diseases which either its excess or deficiency produces, do not of themselves lead to any proper judgment respecting it. On the contrary their fallacious appearance has proved the source of infinite error.

As excitement both depends on exciting powers and excitability, it is evident that when a middle degree of stimulus acts upon a middle degree of excitability, the most perfect effect will be produced. This point, could we ascertain it, might be called the point of health. For the sake of illustration, we may suppose that the greatest excitability of which the living body is capable is 80 degrees: this may be supposed to be the excitability possessed by the body at the commencement of its life, because no part has then been wasted or exhausted by the action of stimuli. Now, if we suppose a scale of excitability to be formed, and divided into 80 equal parts or degrees, the excitability will be wasted or exhausted in proportion to the application of stimuli, from the beginning to the end of the scale. One degree of exciting power applied takes off one degree of excitability, and every subsequent degree impairs the excitability in proportion to its degree of force. Thus a degree of stimulus or exciting power equal to 10 will reduce the excitability to 70, 20 to 60, 30 to 50, 40 to 40, 50 to 30, 60 to 20, 70 to 10, 80 to 0; and, on the contrary, the subtraction of stimulant power will allow the excitability to accumulate.

[DIAGRAM]

The range of good health is ranked from 30 to 50 degrees in the scale; for perfect health, which consists in the middle point only, or at 40 degrees, rarely occurs; in consequence of the variation of the stimuli to which man is continually exposed, such as meat and drink, heat, exercise, and the emotions of the mind, the excitement commonly fluctuates between 30 and 50 degrees, and yet no particular disturbance of the functions takes place. But when at these points, 30 or 50, predisposition commences, the slightest debilitating cause in the former case, and the slightest stimulating cause in the latter, brings on disease, in which the functions begin to be disturbed in various ways, and this disturbance is always in proportion to the hurting powers which have produced the disease, and the delicacy or importance of the part affected.

The effect produced on the excitability by any stimulus, must evidently be in a ratio compounded of the degree of excitability and the force of the stimulus. The same stimulus will produce greater contractions upon a fibre that is more irritable than upon one which possesses less irritability; and the irritability or excitability of the fibre being given, or remaining the same, the contraction will be in proportion to the strength of the stimulus. Hence it is evident, that the effect or excitement must be in a ratio compounded of the exciting powers and excitability.

Sthenic diathesis and disease is caused by the operation of different exciting powers, which produce too great a degree of excitement in the system: this at first increases all the functions, and, when increased, produces a disturbance and inordinate action of them, which is communicated to the whole body. In diseases of this kind there is often an appearance of debility, but this is extremely fallacious, and arises from the disturbed state of the different functions. Hence it is evidently of the utmost consequence to ascertain carefully whether this debility is real, or the effect of asthenic disease: or whether it is owing to the disturbance of the functions by over stimulating, and in this case fallacious; for should a sthenic disease be treated by stimulants and cordials, the effect would be an aggravation of all the symptoms, and a much higher degree of disease.

Asthenic diathesis and disease is brought on by the excitement of the system being diminished: and this may proceed either from a diminution of common stimulant powers, while the excitability is sufficiently abundant, or it may proceed from an exhausted excitability, while the stimulus is sufficiently abundant. The former is called direct, and the latter indirect debility. The exciting causes therefore of asthenic disease, first impair the functions, then occasion a disturbed or inordinate action of them, giving many of them a false appearance; some of them, for instance, appear to be increased, for in hysteria and epilepsy, which are both diseases of debility, the action of the muscles seems to be preternaturally increased; but this depends chiefly on the accumulated excitability, which gives such a degree of irritability to the system, that the smallest irritation, whether external, such as heat, exercise, &c. or internal, as emotions of the mind, excite a strong spasmodic action, which brings on the symptoms of epilepsy and hysteria. This inordinate action however soon exhausts the morbid excitability, and thus suspends itself, a sleep often follows, from which the patient wakes with only a general sense of languor and debility: but as the same cause still remains, the excitability of the body again becomes morbidly accumulated, and thus the slightest stimulus produces a recurrence of the fit, and the tendency to return will increase with its recurrence, so that at last the slightest imaginable cause will produce it, on account of the power of habit and association.

Gout likewise appears like a sthenic disease, and in inflammation takes place, which resembles pleurisy or peripneumony; but this symptom is fallacious, for it depends on debility, and is only to be cured by means, which in pleurisy and peripneumony, would produce death.

Hence it must be evident that those phenomena of diseases, which we call symptoms, are generally fallacious; but this may be owing to our imperfect knowledge of the animal economy, so that we are not able to explain or understand the manner in which they are produced: we ought however carefully to guard against being misled by them in practice. The great difficulty is to distinguish the nature of the disease, whether it is sthenic or asthenic, or whether it depends on too great excitement, or on debility; for this being once clearly ascertained, we proceed with certainty in our mode of treatment, instead of the random practice, which must be the consequence of not taking a proper view of the laws of life, and the causes of diseases.

The nature of the disease may be generally ascertained, by attending to the habits of the patient, and the manner in which he has lived, as well as to the state of the pulse; but in cases where these circumstances do not render it clear, it may be ascertained, beyond a doubt, by a trifling degree of stimulus, as, for instance, by any cordial, as a little wine or spirits. If the disease be of an inflammatory or sthenic kind, the symptoms will be aggravated, and the cordial will not produce its usual pleasant effects on the system; but on the contrary, if the nature of the disease be astheric, then the usual pleasant effects of the cordial will be perceived, and the pain and other symptoms will be alleviated. This trial, which is soon made, and without danger, will determine our plan of cure, and we can then proceed with the most perfect certainty. Thus you will see that this view we have taken leads to a very different and much more rational plan of practice than is generally followed, in which the most judicious physicians confess that they have no clue to guide them; and complain that the science of medicine consists merely in a number of insulated facts, not connected by any theory: that they merely prescribe a remedy because they have seen it of use in an apparently similar state, but that they have no certainty of its producing a similar effect in the cases in which they prescribe it. This all depends on trusting to the fallacious appearance of symptoms, and not having taken a proper view of the laws of life, or the manner in which the exciting powers act on living bodies.

After these observations on the diagnosis, or the method of distinguishing the nature of diseases, I shall proceed to consider more particularly the nature of sthenic diseases, and the methods of curing them, which will occupy the remainder of our time this evening.

The powers or causes, which by their action produce inflammatory or sthenic diseases, are, first, heat, which is a very frequent cause, particularly when it succeeds cold; for the cold accumulates the excitability, and then renders the whole body, or a part, more liable to be affected by the heat afterwards applied. In this way is produced rheumatism, catarrh, or, as it is commonly called, a cold, and peripneumony. These complaints have been often attributed to cold, but I believe that there never was a well attested instance where cold alone, without being either followed by heat or some other stimulus, produced a real sthenic, or inflammatory disease. This is not merely a distinction, it is a circumstance of the utmost importance, because it influences the mode of practice to be pursued. Heat is one of the exciting or stimulant powers which support life, and one of the most powerful of these stimulants; but cold is only a diminution of it: how then can this produce a sthenic state, or a state of too high excitement? The blood is one of the exciting powers, which, by its continual circulation supports life; but surely if we abstracted a quantity of this fluid from the body, no person will be bold enough to say, that we by that means should produce an inflammatory disease. Cold renders the body more liable to be affected by heat, or any other stimulus applied, but does not of itself produce any stimulant or inflammatory effects.

To see more clearly the manner in which cold acts, let us inquire how it produces or contributes towards the production of catarrh. When we go into the cold air, at every respiration we take a quantity of it into the lungs, which brushes over the surface of the mucous membrane that lines the nostrils and trachea, and thus, robbing them of their heat, allows the excitability to accumulate. But we feel no fever, no sense of tightness or stuffing, nor any other symptom of catarrh, so long as we continue in the cold. If however we afterwards go into a warm room, and particularly near a fire, we receive by the act of respiration the warm air into those very parts which have been previously exposed to cold, and whose excitability is consequently accumulated. The first effect we perceive is a glow of the parts, which is by no means unpleasant, this however increases; and, in the course of half an hour or an hour, a sense of dryness and huskiness comes on, with a sensation of stuffing in the nostrils, and a tendency to a short dry cough: often likewise, if the exposure to cold has been considerable, and the heat afterwards applied great and sudden, we experience a shivering, and other symptoms of fever. These symptoms are all increased by taking into the stomach any liquid that is either of warm temperature or stimulating quality, or particularly both; we spend a restless night, and awake with all the symptoms of a catarrh, or cold, as it is improperly called. For it is evidently an inflammatory fever, and can be speedily cured by the debilitating plan, and particularly by keeping in a moderately cool place, where the temperature is equable, and not subject to alternations of heat and cold.

But how easily might this complaint have been avoided, were the person subject to it acquainted with its real nature, and the manner in which it is brought on. When we come out of a very cold atmosphere, we should not at first go into a room that has a fire in it; or, if this cannot be well avoided, we should keep for a considerable time at as great a distance from the fire as possible, that the accumulated excitability may be gradually exhausted by the moderate and gentle action of heat; and then we may bear the heat of the fire without any danger; but above all, we should refrain from taking warm or strong liquors while we are hot. In confirmation of this opinion, numerous instances might be brought, where catarrh was cured merely by exposure to cold.

When a part of the body only has been exposed to the action of cold, and the rest kept heated; if, for instance, a person in a warm room has been sitting so that a current of air, coming through a broken window, has fallen upon any part of the body, that part will soon be affected with an inflammation, or what is called a rheumatic affection. In this case, the excitability of the part exposed to the action of the cold, becomes accumulated, and the warm blood, rushing through it, from every other part of the body, excites an inflammation.

Thus catarrh and rheumatism are inflammatory complaints, or depend on too great a degree of excitement, and are to be cured by lowering the excitement, or diminishing the action of the exciting powers; by bleeding, purging, low diet, and particularly keeping in a moderately cool place; and these complaints will be as speedily and certainly cured by these methods, properly and judiciously persevered in, as a slight cut or wound will be healed by what surgeons call the first intention.

There are complaints which resemble these, but whose nature, however, is very different, and which require a very different mode of treatment. After a part has been long affected with rheumatic inflammation the excitability of the muscular fibres becomes so far exhausted, that a state of indirect debility takes place, and an inflammation, accompanied with pain and redness, which is very different from that I formerly described, as it depends upon a debilitated or relaxed state of the parts, instead of too great a degree of excitement. This instance shows strongly the fallacy of symptoms; but it may be readily distinguished from the inflammatory rheumatism, by attention to the effects of the exciting causes. The inflammatory rheumatism is aggravated by heat, hence it is more violent in bed than at any other time. The latter complaint, however, is greatly relieved by heat: the warm bath alleviates all the symptoms; so does a warm bed. It is evident that these diseases, though attended by the same symptoms, are as opposite, and require as different modes of treatment as an inflammation of the brain, and a dropsy. The inflammatory state has been called the acute rheumatism, and the other, the chronic rheumatism; I would, however, prefer the terms sthenic and asthenic rheumatism.

In the same manner, there is a catarrh, which is liable to afflict persons who have often been subject to an inflammatory cold, particularly persons advanced in years; and this depends on a state of indirect debility of the parts, the excitability of which has been exhausted by frequent and violent inflammatory affections. This complaint, which I would call asthenic catarrh, requires directly opposite treatment from the inflammatory or sthenic catarrh. The latter is aggravated by heat, but relieved by a cool temperature. Warm air is peculiarly grateful to those who are afflicted with the former, and if they go into a cool temperature, they are immediately seized with cough, and expectoration; for the disease being a disease of debility, the withdrawing the stimulus of heat, must increase it. The excitability of the parts is so far exhausted, that it requires a stimulus even more than natural to keep them in tone: hence persons labouring under asthenic catarrh, and some species of asthma, which are only varieties of this disease, find themselves best when exposed to a warm temperature, but on the heat being diminished, and consequently the parts relaxed, the cough and difficulty of breathing immediately come on.

Having examined the effect of heat, in producing inflammatory or sthenic disease, I now proceed to the consideration of the other powers. Of the articles of diet, the only food in danger of being too stimulant, is perhaps flesh or land animal food, used in too great quantity, particularly when seasoned, a preparation which adds much to its stimulant power. Spirituous and vinous liquors, let them be ever so weak or much diluted, stimulate more quickly, and more readily than seasoned food, and their stimulus is in proportion to the quantity of alcohol which they contain These substances, when conjoined with rich food, must bring on a predisposition to sthenic disease, in almost any constitution, particularly in the young and healthy, and, in many instances, those diseases actually take place; or should this not be the case, should the person avoid, or escape the effects of inflammatory diseases, the excitability will be exhausted, and diseases of indirect debility, such as gout, apoplexy, indigestion, palsy, &c. will take place.

These stimulants are never necessary to a good constitution, and their effects will always, sooner or later, be experienced: for though a person with a good constitution may continue for years to indulge in the pleasures of the bottle, or the luxuries of the table, depend upon it that a continuance of them will sap the vigour of the strongest constitution that ever existed.

As nothing contributes more to the health of the body than moderate and frequently repeated exercise, which rouses the muscles to contraction, and promotes the circulation of the blood in the veins towards the heart: it thus produces excitement; but an excess of it will produce sthenic diathesis; and, if carried to great excess, it will produce a state of indirect debility, or exhausted excitability.

When any, or all of these exciting powers act too strongly on the body, the first effect they produce is a preternatural acuteness of all the senses; the motions, both voluntary and involuntary, are performed with vigour, and there is an acuteness of genius and intellectual power. In short, every part of the body seems in a state of complete vigour and strength; that this is the case with the heart and arteries, appears from the strong and firm pulse; in the stomach it is shown by the appetite; and, in the extreme parts, by the ruddy colour and complexion. In short, every appearance marks vigour of the body, and abundance of blood. Could the body be kept in this state, nothing could be more to be desired; this, however, is impossible; the excitement, though still within the bounds of health, has overstepped the point of good health, and is verging fast to predisposition to sthenic disease; so that, to secure a permanent state of health, it is always better to keep the excitement rather under the middle point, or 40 degrees, than above it. During the predisposition to sthenic disease, which is produced by the longer continued, or increased action of these powers, no symptoms of disease appear; but shortly after, disturbed sleep, depressed spirits, languor, a sense of fulness, heaviness, particularly after eating, show that this sthenic state cannot be further increased with impunity. The least increase of sthenic diathesis now brings on a disturbance of the functions, or actual disease; the commencement of which is generally a shivering, and a sense of cold; thirst and heat succeed; and then generally a pain in some part, either external or internal: costiveness generally attends this state, the urine is clear, and secreted in small quantity;

memory and imagination become diminished, and there is generally less appetite for food.

In peripneumony, inflammatory sore throat, and acute rheumatism, there is an inflamed condition of the lungs, of the parts about the throat, or of the muscles of the extremities: this shows that the excitement here is greater than in other parts of the body; but it is still increased or too great in every part, only those parts which give the peculiar character to the disease are more affected than other parts of the body, by being more exposed to the exciting causes: thus, if a person be in perfect health, or a little below, he will not be easily affected by any of the exciting causes of sthenic disease, unless their application be very violent; he will go into a warm room out of the cold air, and feel no other effect than a pleasant glow: but if, by high living, or other means, he is brought near the point of predisposition to sthenic disease, then the slightest additional stimulus will bring it on, and if the throat has been exposed to the application of cold, and the person comes afterwards into a heated room, an inflammation of the parts about the throat, or an inflammatory sore throat, accompanied by a sthenic diathesis of the whole system, will be the consequence. This cannot be cured by merely diminishing the excitement of the part, while the excitement of the whole system remains: if we apply leeches to the throat in this state, to diminish the quantity of blood, we only debilitate the vessels, while fresh quantities of blood are poured into them from the too full vessels of the body; even if we could thus remove the sthenic diathesis of the part, we should go but a little way towards removing the inflammatory disease, which universally pervades the system.

The mode to be pursued therefore is, to take a quantity of blood from the body, by opening a vein; to keep the body cool, by remaining in a room where the temperature is at temperate, or a little below; by abstaining from animal food, and from spirituous or fermented liquors; and by the exhibition of purgatives, or at least of laxatives. Then leeches or blisters applied to the part affected will produce a good effect; and even stimulant applications to the inflamed part may be advantageous; for a topical inflammation, as we shall afterwards have occasion to see, depends on a debilitated state of the minute vessels of the part, while at the same time the action of the whole system is increased.

Besides the energy of the exciting hurtful powers, which I have mentioned, there is in the parts which undergo the inflammation, a greater sensibility, or an accumulated excitability; by which it happens that some are more affected than the rest. To this we may add, that whatsoever part may have been injured by inflammation, that part in every future sthenic attack is in more danger of being inflamed than the rest. Hence inflammatory sore throat, rheumatism, and some other complaints of the kind, when once they have supervened, are very apt to recur.

Among the sthenic or inflammatory diseases may be enumerated rheumatism, catarrh, cynanche, or sore throat, scarlet fever, inflammations of the brain, stomach, lungs, &c. &c.

Many of the contagious diseases, particularly small pox and measles, produce a sthenic state, and are to be cured, or their action moderated, by the debilitating plan which has been pointed out; and particularly by a moderate, constant, and equable diminution of temperature. Hence the violence of these diseases is greater when they attack a person already predisposed to sthenic diathesis, but much more mild when the excitement is rather under par.

LECTURE XII. ON INFLAMMATION AND ASTHENIC DISEASES.

The last lecture was taken up chiefly with an account of sthenic diseases, or those depending on too great a degree of excitement, and which have been generally, but improperly, called inflammatory or phlogistic. In that lecture I attempted to show, that when the natural exciting powers, which support life, act with too much power, or particularly if we employ any stimulants not natural to the body, the functions both of body and mind become increased in vigour; but if the exciting causes are continued and increased, the functions become disturbed, and their action becomes painful and distressing. This state, which is called sthenic diathesis, is often accompanied by a redness, swelling, pain, and increased heat of some particular part: these symptoms constitute what is usually termed an inflammation of the part.

The method of cure in sthenic diseases was shown to be, by reducing or moderating the action of the exciting powers; by keeping the body cool; abstaining from high seasoned, and, in general, from animal food; by the use of purgatives, and in many cases by diminishing the quantity of blood in the body. I mentioned likewise, that it would be but of little use to attempt to subdue the excitement of the inflamed part, unless the excitement of the whole system was previously diminished; but that after a general bloodletting, stimulant remedies applied to the inflamed part, might be employed with success. This is strictly agreeable to experience, but at first sight seems so very contrary to the principles that have been advanced, that I shall endeavour to explain the phenomena of inflammation, which do not seem to be in general well understood.

All kinds of inflammation agree, in being attended with redness, increased temperature, pain, and swelling; but they vary according to the situation and texture of the part affected. All parts of the body, excepting the cuticle, nails, hardest part of the teeth, and hair, are subject to inflammation.

Among the causes of these complaints, may be enumerated too full a diet, particularly too free a use of fermented liquors, and whatever increases the impetus of the blood towards the part, as mechanical and chemical irritation, and sudden changes of temperature, particularly from cold to heat.

To explain the nature of inflammation, it may be observed, that such is the wise constitution of the animal body, that whatever injures it, excites motions calculated to correct or expel the offending cause. Thus if an irritating substance is received into the stomach, it excites vomiting; if into the lungs, a violent fit of coughing is excited, and if into the nostrils, sneezing is the consequence. In such cases we can readily trace the motions excited, and the manner in which they act; but cannot trace the manner in which the offending cause excites these motions.

Now if it can be shown that inflammation, like vomiting and coughing, is an effort of the system to remove an offending cause, and if we can trace every step of this operation, with the exception of the changes induced on the nervous system, we shall understand the nature of inflammation as completely as that of any function of the body.

The circumstance the most difficult to explain, is the increased redness of the part affected, which can only depend on an increased quantity of blood in the vessels. This has been supposed to depend upon an increased action of the vessels of the part; but that this is not the case, must be evident from what was said when we were speaking of the circulation of the blood. It was shown, that the circulation could not be carried on by the mere force of elasticity alone; this force, were it perfect, would produce no effect; but as there is no body with which we are acquainted that is perfectly elastic; so the coats of the arteries are very far from being so, hence their effect as elastic tubes will be to diminish the force of the heart, instead of adding to it; for a certain quantity of this force will be spent in distending the vessels, which, were they perfectly elastic, would be restored to them, but as this is not the case, this force is by no means restored. Indeed a variety of considerations, observations, and experiments, tend to prove, that the vessels are endowed with a power very different from elasticity, which differs only in degree from that of the heart; in short, they are possessed of muscular power.

After each contraction of the muscular coat, the elastic will act as its antagonist, and enlarge the diameter, till the vessel arrive at a mean degree of dilatation, but after this there is no further power of distention inherent in the vessel. The action of the elastic coat ceases; and no one will assert that a muscular fibre has power to distend itself.

The only power by which the vessel can be further distended, is the vis a tergo: after the vessel arrives at its mean degree of dilatation, both the elastic and muscular coats act as antagonists to the vis a tergo, or force which propels the blood into, and thus tends further to dilate the vessel. If then the vis a tergo become greater than in health, the powers of resistance inherent in the vessels remaining the same; or if the latter be weakened, the vis a tergo, or propelling force, remaining the same, the vessel must suffer a morbid degree of dilatation. These appear to be the only circumstances under which a vessel can suffer such dilatation.

But if, while the powers of the vessels remain the same, the vis a tergo, or propelling force, be diminished, or the propelling force remaining the same, the power of the vessels become increased; then an opposite condition or state of the vessels, viz. a preternatural diminution of their area, will take place.

In the one case the distending force bears too great a proportion to the resisting force; and preternatural distention is the consequence. In the other the resisting force bears too great a proportion to the distending force, and preternatural contraction is the consequence.

It is not necessary that the vessels should be in a state of greater debility than in health, in order that an inflammation or distention may take place: it is only necessary that the proportion which their action bears to the propelling force be less than in health. If the propelling force remain the same, the vessels must be in a state of debility before an inflammation can take place; but if the propelling force be increased by a fullness of the vessels and sthenic diathesis, inflammation may take place, although the vessels of the part act as powerfully as in health, or more so. But after inflammation has taken place, as the vessels are preternaturally distended, they must also be debilitated.

The degree of inflammation is not however proportioned to the debility of the minute vessels of an inflamed part, but to the diminished proportion of their power to the propelling force.

When, therefore, inflammation arises from an increased action of the arterial system, or an increased propelling force, while the force of the capillaries or minute vessels remains the same, it constitutes what is called an active inflammation, and is to be cured by general bleedings, and then by gentle applications of tonics to the part, to increase its action; but when it arises from a debility of the minute vessels, without any increase of the propelling force, it forms what is known by the name of passive inflammation; in which general bleeding is not required, but the application of stimulants and tonics to the inflamed part to enable the vessels to recover their lost tone, and restore the balance between their action and the vis a tergo. From what has been said, it must be evident, that if inflammation depend on the diminished proportion of the power of the capillaries to the propelling force, it will be more apt to supervene under the three following circumstances.

1. In a state of plethora, because then all the vessels are over distended, and consequently any cause tending further to distend

them, whether it be a cause which debilitates them, or increases the propelling force, will be more felt than in health.

2. In a state of general debility, because the vital powers in any part are more readily destroyed than in health.

3. In a state of general excitement, because then the propelling force is every where strong, and consequently apt to occasion distention of the vessels, wherever any degree of debility occurs. These are the states of the system which are found to predispose to inflammation. In the first and last, the inflammation is generally of that kind, which is termed active: the propelling force is considerable, and the larger arteries are readily excited to increased action. In the second state the inflammation is of the passive kind.

This is not merely a useless physiological disquisition; it is of the greatest use in directing our practice; and teaches us that, in passive inflammation, which has all the symptoms of active, and therefore shows in a striking point of view the fallacy of symptoms, we shall not succeed by applying leeches, and other debilitating means, to the inflamed part; on the contrary, we shall aggravate the complaint; and the cure must be effected by stimulants applied to the part.

As an instance of this kind of inflammation, I may mention that kind of ophthalmia or inflammation of the eyes, which is of long standing, and which not only resists the powers of leeches and blisters, but is increased by them. I have frequently been consulted by patients, who had for months been under the debilitating plan, without any benefit; and who have been relieved almost instantly by the application of electricity and a stimulating lotion, which restored the tone of the debilitated vessels of the sclerotic coat, and enabled them to expel their overcharged contents; and the balance between their action and the propelling force being restored, the inflammation disappeared.

Indeed the effects of electricity in these kinds of inflammations are wonderful: it seems to act almost by a charm, so quickly does the inflammation subside; but when we understand the nature of this kind of inflammation, it is nothing but what we might expect from its action.

I have been thus minute on the subject of inflammation, because the theory of it, which I have attempted to defend, differs considerably from the commonly received opinions. I shall now proceed to consider the nature of asthenic diseases.

From what has been already said, it must be evident that the causes of diseases which we have assigned, are very different from those delivered by physicians who preceded Dr. Brown. Some physicians imagined that diseases were caused by a change in the qualities of the fluids, which became sometimes acid, and sometimes alkaline; or on a change of figure of the particles of the blood: some imagined diseases to be owing to a rational principle, which they called the vis medicatrix naturae, which governed the actions of the body, and excited fever or commotion in the system to remove any hurtful cause, or expel any morbid matter, which might have insinuated itself into the body. Others supposed many diseases to arise from a constriction of the extreme vessels by cold; or from a spasm of them, which was a contrivance of the vis medicatrix, to rouse the action of the heart and arteries to remove the debility induced.

We have seen, however, that health and diseases are the same state, and depending upon the same cause; viz. excitement, but differing in degree; and that the powers producing both are the same, sometimes acting with a proper degree of force: at other times either with too much, or too little.

We shall now examine how the diminished actions of the different exciting powers produce asthenic disease; and we shall take them in the same order as when we were speaking of sthenic diseases. It must be recollected however that an asthenic state, or a state of debility, may be produced in two ways. First, by directly diminishing the action of the exciting powers. Secondly, by exhausting the excitability, by a strong or long continued stimulant action. The former state is called direct debility, and the latter indirect debility. This is not merely a distinction without a difference, the body is in very different states, under these two different forms of disease. In the former case, the excitability is abundant, and highly susceptible of the action of stimulants. In the latter, it is exhausted, and the body has very little susceptibility.

Cold, or a diminution of heat, carried beyond a certain degree, is unfriendly to all animals. Dr. Beddoes has shown very clearly in his Hygeia, that it is the cause of a great many diseases which take place at boarding schools, and that it there gives origin to a great number of diseases that afterwards arise, and, indeed, not unfrequently ruins the constitution. It produces relaxation of the vessels, asthenic or passive inflammation, and even gangrene. He has shown that in most schools children are afflicted with chilblains from this cause; this is a case of passive inflammation, but is only a symptom of the general debility induced, which shows itself afterwards by the production of other symptoms. Hence it is necessary for the preservation of health, that the temperature of school rooms should always be kept equable, and regulated by means of a thermometer. It should not exceed 50 degrees, nor should it be allowed to fall much below it. If precautions of this kind are thought to be necessary, and practised with uncommon attention, in places where vegetables are reared, surely they ought not to be neglected in those seminaries where the human species are to be brought to maturity, and a good constitution established.

But though I have no doubt whatever, that this equable temperature would prevent a number of diseases, which originate in too low a temperature, yet I am far from wishing to have it thought that I would not induce a hardy state of the constitution, which would enable it to bear the vicissitudes to which it must be exposed in its journey through life, by every means in my power. Hardiness is the most enviable of all the attributes of animal nature, and can neither be acquired, nor recovered when it is lost, but upon certain terms, to which many people submit with reluctance, because they must give up many indulgences and gratifications with which it is utterly inconsistent.

One of the causes that chiefly contributes to reduce persons living in affluence below the standard of hardiness, is the dependence they place on a considerable degree of external warmth, for preserving a comfortable state of sensation. From what has been said again and again in some of the latest of these lectures, it must be evident that continued warmth renders the living system less capable of being excited to strong, healthy, and pleasant action: heat in excess, whether it may be excess of duration or intensity, constantly debilitates, by exhausting the excitability of the system, and thus producing a state of indirect debility. Every muscle steeped in a heated medium, whether of air or water, loses much of its contractibility. A heart kept in heated air, or put in hot water, will not contract on the application of a stimulus; even the limb of a frog, when heated in this manner, ceases to move on the application of the galvanic exciters. Every nerve grows languid, and when it does become excited, it acquires a disposition to throw the moving fibres, with which it is connected, into starts, twitchings, and other irregular convulsive motions. Though therefore nothing can more contribute to the health of the body than a moderate and well regulated temperature, about 48 or 50 degrees, sometimes for a short interval a little lower, when exercise is taken at the same time, yet when we consider the life led by persons of fashion, we should hope that it proceeded from ignorance of these consequences; so diametrically opposite is it to the dictates of nature and reason.

Instead of rising from table after dinner, and availing themselves of the cooling and refreshing qualities of the air, even in the finest seasons, when every thing which pure and simple nature can offer, invites them abroad, they do every thing they can, as Dr. Beddoes observes, to add to the overstimulating operation of a full and hearty dinner. After taking strong wine with their food, they sit in rooms rendered progressively warmer, all the afternoon, by the presence of company, by the increase of fires, and for more than half the year, by the early closing of the shutters, and letting down of the window curtains. After a short interval, tea and coffee succeed; liquors stimulating both by their inherent qualities, and by virtue of the temperature at which they are often drank. And that nothing may be wanting to their pernicious effect, they are frequently taken in the very stew and squeeze of a fashionable mob. The season of sleep succeeds, and to crown the adventures of the evening, the bed room is fastened close, and made stifling by a fire: and though the robust may not quickly feel the effects of this mode of life, with the feeble it is quite otherwise. These, as they usually manage, rarely pass a few hours of sleep without feverishness and uneasy dreams; both of which contribute to their finding themselves by far more spent and spiritless in the morning, than after their evening fit of forced excitement, instead of having their spirits and strength recruited by the "chief nourisher in life's feast," Perhaps they drink tea before rising, and indulge in a morning nap; this weakens much more than the greatest muscular exertion they would be capable of supporting for an equal time. For the sleep at this time is almost invariably disturbed, and attended by a heat of the skin. The reason of this must be evident to every one who has attended these lectures.

The effect of sleep is to accumulate the excitability, or render it more sensible to the effects of any stimulants applied. This takes place in every constitution, and much more in the more delicate: hence the heat of the bed, and of the tea, acts so powerfully on the surface, as, in general, to produce great perspiration, or, at any rate, great languor and debility.

Let me ask, can any one, who lives in this manner, expect to enjoy good health? With as great probability might we expect, that when we plunged a thermometer into hot water, the mercury would not rise, or when we applied a lighted match to gunpowder, it would not explode. The laws of nature are constant and uniform, and the same, or similar causes, both in the animate and inanimate world, are always productive of the same, or similar effects.

The cure of these complaints is at least obvious, if not easy. It consists in deserting crowded and heated rooms, at least for part of the time they have been usually occupied; in abstaining from strong wines; in keeping the bed rooms moderately cool; and retiring to rest at a proper hour.

With respect to the effects of nutriment, in producing asthenic diseases, we may observe, that all watery vegetable food, too sparing a use of animal food, as also meat which is too salt, and deprived of its nutritious juices by keeping, when more nutritious matter is at the same time withheld, constantly weaken, and thereby tend to produce asthenic diseases. Hence would appear to arise that remarkable imbecility of body and mind which distinguishes the Gentoos. Hence arise the diseases with which the poor are every where afflicted; hence scrofula, epilepsy, and the whole band of asthenic diseases.

But intemperance in eating and drinking, or taking nutritious and highly stimulant substances too freely, will, infallibly, bring on asthenic disease, or a state of indirect debility, by exhausting the excitability; and it must be observed, that this species of debility is much worse to cure than the direct kind; for in the latter we have abundance of excitability, and a variety of stimuli, by which we can exhaust it to the proper degree, and thus bring about the healthy state; whereas, in indirect debility, the vital principle or excitability is deficient; and we have not the means of reproducing it, at pleasure, absolutely under our command. Besides, the subtraction of stimulants, which is one of the most certain means we have of accumulating excitability, if carried to a great extent, in diseases of indirect debility, would produce death, before the system had power to reproduce the lost or exhausted excitability. Hence the cure, in these two kinds of debility, must be very different: in cases of direct debility, as in epilepsy, we must begin with gentle stimulants, and increase them with the greatest caution, till the healthy state is established: we must, however, guard most carefully against over doing it; for, if we should once overstep the bounds of excitement, and convert the direct into indirect debility, we shall have a disease to combat, in which we have both a want of excitement and of exciting power.

In cases of violent indirect debility, as, for instance, in gout, when it affects the stomach: it would be wrong to withdraw the stimulus, for the excitability is in such an exhausted state as to produce no action, or very imperfect and diseased, from the effect of the common exciting powers; we must, therefore, here apply a stimulus greater than natural, to bring on a vigorous and healthy action, and this stimulus we should gradually diminish, in order to allow the excitability to accumulate, by which the healthy state will be gradually restored.

This method was very judiciously recommended, by a very eminent physician, in the case of a Highland chieftain, who had brought on dreadful symptoms of indigestion by the use of whisky, of which he drank a large silver cup full five or six times in the day. The doctor did not merely say, diminish the quantity of spirits gradually, for that simple advice would not have been followed; but he advised him to drink the cup the same number of times full, but each morning to melt into it as much wax as would receive the impression of the family seal. This direction, which had something magical in it in the mind of the chieftain, was punctually obeyed. In a few months the cup was filled with wax, and would hold no more spirits; but it had thus been gradually diminished, and the patient was cured.

This reminds me of a number of cases, which had been brought on by drinking porter, and other stimulant liquors, without knowing the taste of water. In many of these cases if a moderate quantity of water were drank every day they would be cured; but you would find few who would follow such plain and simple directions. How then must a physician proceed? Why, as is generally done by the most judicious: they direct their patients to Bath or Buxton, and there advise them to swallow a certain quantity of water every day, which they do most scrupulously, and, of course, return home cured.

As causes of asthenic disease, we must not omit the undue exercise of the intellectual functions. Thinking is a powerful exciting cause, and produces effects similar to those of intoxication. None of the exciting powers have more influence upon our activity, than the exercise of the intellectual powers, as well as passion and emotion. Homer, the great observer and copyist of nature, observes of the hero, whom he gives for a pattern of eloquence, that, upon his first address, before he had got into his train of thought, he was awkward in every motion, and in his whole attitude; he looked down upon the ground, and his hands hung straight along his sides, as if they had lost the power of motion; and his whole appearance was a picture of torpidity. But when he had once fairly entered upon his subject, his eyes were all on fire, his limbs all motion, grace, and energy.

Hence, as the exercise of the intellectual functions evidently stimulates, an excess of thinking must bring on indirect debility, by exhausting the excitability. But though we do meet with instances of indirect debility arising from this source, it must be confessed that they much oftener arise from the use of very different stimulants.

As excessive exercise of the intellectual powers will bring on indirect debility, so the deficient, weak, or vacant state of mind, which is unable to carry on a train of thinking, will produce direct debility. Indeed this debility often occurs to those whose minds have been all their life actively engaged in business, but who have at last retired to enjoy themselves, without having a cultivated mind fit for retirement. They become languid, inert, and low spirited, for want of the stimulus of mental exertion; and in many cases cannot be completely restored to health, till they are again engaged in their usual occupations.

Violent passions of the mind, such as great anger, keen grief, or immoderate joy, often go to such an extent as to exhaust the excitability, and bring on diseases of indirect debility. Hence both epilepsy and apoplexy have been the consequences of violent passion.

On the contrary, when there is a deficiency of exciting passion, as in melancholy, fear, despair, &c. which are only lower degrees or diminutions of joy, assurance, and hope, in the same way that cold is a diminution of heat, this produces a state of direct debility. The immediate consequences observable are, loss of appetite, loathing of food, sickness of the stomach, vomiting, pain of the stomach, colic, and even low fevers.

The effect of impure air, or air containing too small a proportion of oxygen, is likewise a very powerful cause of debility.

In short, when any or several of these causes, which have been mentioned, act upon the body, asthenic diseases are the consequence.

Asthenic diseases, as has frequently been hinted, may be divided into two classes, those of direct debility, and those of indirect debility.

Among the diseases of direct debility may be enumerated dyspepsia, hypochondriasis, hysteric complaints, epilepsy, bleeding of the nose, spitting and other effusions of blood, cholera morbus, chorea, rickets, scrofula, scurvy, diabetes, dropsy, worms, diarrhoea, asthma, cramp, intermittent fevers.

Among those of indirect debility, or which are produced by over stimulating, which exhausts the excitability, may be enumerated, gout, apoplexy, palsy, jaundice, and chronic inflammation of the liver, violent indigestion, confluent small pox, typhous fever, and probably the plague, dysentery, putrid sore throat, tetanus.

Diseases, therefore, according to this system, may be divided into two classes. First, general diseases, which commence with an affection of the whole system, and which must be accounted general, though some part may be more affected than the rest. Secondly, local diseases, which originate in a part, and which are to be regarded as local, though they may sometimes in their progress affect the whole system, like universal diseases: still however they are to be cured by remedies, applied not to the whole system, but to the part affected only.

A pleurisy or peripneumony, for instance, is a general disease, though the chief seat of the symptoms seems confined to a portion of the thorax: but the affection of this part, though it may be somewhat greater than that of any other equal part, is vastly less than the affection or diathesis diffused over the whole body. The exciting or hurtful causes which produce these diseases, by no means exert their whole power upon a small portion of the superficial vessels of the lungs, and leave the rest untouched; on the contrary, they affect exery part of the system, and the whole body partakes of the morbid change. Indeed the general or universal affection; viz. a sense of heaviness and fullness, uneasy sleep, and other symptoms of increased excitement, are commonly perceived some time before the pain of the thorax becomes sensible. The remedies which remove the disease; viz. venesection, abstaining from animal food, and every mode of debilitating, do not exert their whole efficacy on an inflamed portion of the lungs; for by removing the affection of the lungs we should go but a little way towards removing the disease.

Among local diseases we may enumerate wounds, or solutions of the continuity of the part, bruises, fractures, inflammations from local irritation, &c. Hence it is evident that the treatment of general diseases is the province of the physician; and of local ones of the surgeon. But there are some general diseases which are apt to degenerate into local, and therefore require the attention both of the physician and the surgeon. Among these we may reckon suppuration and gangrene, sphacelus, and some others.

The first class, or general diseases, may be divided into two orders; sthenic, and asthenic. The asthenic order may be subdivided into two genera; viz. diseases of direct debility, and diseases of indirect debility; for debility, according to the system I am explaining, is that relaxed or atonic state of the system which accompanies a deficient action of the stimulant or exciting powers; and this deficient action may arise immediately from the partial or too sparing application of the exciting powers; the excitability or capacity of the system to receive their actions, being unaffected or sufficiently abundant; or it may arise from the excitability being exhausted, by the violent or long continued action of the exciting powers.

This arrangement of diseases, which naturally follows from the fundamental principles of the doctrine, and which is guided by the state and degree of excitement, is widely different from that of former nosologists, who have arranged or classed them according to symptoms, which have already been shown to be fallacious; and which method of arrangement brings together diseases the most opposite in their nature, and separates those most nearly allied. This is evident in every part of the nosology of Sauvages and Cullen. In the genus cynanche of the latter, are placed the common sthenic or inflammatory sore throat, or cynanche tonsillaris, and the putrid or gangrenous sore throat, the cynanche maligna: the former is a sthenic disease; the latter one of the greatest debility; yet they have the same generic name.

The mode of classing diseases which I have adopted, after the example of Dr. Brown, is the consequence of first taking a view of the nature of life, and the manner in which it is supported; and from thence observing how those variations from the healthy state, called diseases, are produced; and this is certainly the proper plan; for, as every effect will be produced with more accuracy, whilst its cause is acting in a proper degree, it is certainly right to begin by drawing our general propositions from the healthy state; by which means we avoid being misled by those false appearances which the living system puts on, during a morbid state; and though the contrary has been the general practice of nosologists and pathologists, I must confess it appears to me like beginning where the end should be; for to lay down rules for restoring health, and begin by observing the phenomena of disease, is like building a house, and beginning with the roof.

In the last lecture I pointed out the general method of curing sthenic diseases; I shall now proceed to the cure of asthenic, and shall begin with those depending on direct debility, as in these diseases the excitability is morbidly accumulated, and consequently more liable to be overpowered by the action of a stimulus, we must, therefore, at first, apply very gentle stimulants, increasing them by degrees, till the excitement be arrived at the healthy state.

In cases of indirect debility, the excitability is so far exhausted as not to be sufficiently acted on by the ordinary powers which support life; we must therefore employ, at first, pretty strong stimulants, to keep up such a degree of action as is necessary to preserve life; we should, however, be careful not to overdo it; for our intention here, in giving these stimuli, is only to keep up life, while the cure must depend upon the accumulation of the excitability. That this may take place, therefore, we must gradually lessen the quantity of stimulus, till the excitability become capable of being sufficiently acted on by the exciting powers, when the cure will be affected.

There is, however, an important point, with respect to the cure of diseases of exhausted excitability, which could not be known to Dr. Brown; and this depends on the fact which was formerly pointed out; viz. that the degree of excitability was in proportion to the oxydation of the system. On this account I have given the oxygenated muriate of potash in typhus, which is a disease of diminished excitability, in more than one hundred cases, without the loss of one, a success which has attended no other mode of practice in this disease, if we except, perhaps, the affusion of cold water, as described by Dr. Currie, the effects of which are wonderful, but which can only be applied at the commencement of the disease. In all diseases of indirect debility, therefore, it is proper to attempt the introduction of oxygen into the system, by the oxygenated muriate of potash, acid fruits, nitre, &c. I do not think that the inhaling of oxygen gas for a few minutes in the day can do much good; but free ventilation of apartments, and gentle exercise in the open air, are highly useful.

In either case of debility, we should by no means rely on the action of medicines alone; for though there are a variety of stimulants which will produce excitement, yet this is only temporary, we must therefore endeavour, by nutritious substances, to fill the vessels with blood, and employ all the natural exciting powers in due proportion as soon as possible.

But in the cure of either sthenic or asthenic diseases we shall seldom succeed by the use of one remedy only: for since no stimulus exerts its effects equally on all parts of the body, but always acts more powerfully on some part than on others, we cannot by the use of one remedy alone obtain an equal increase or diminution of excitement.

There are few diseases however in which the excitement is equally increased or diminished over the body; some part being generally more affected than the rest; and this inequality produces the various phenomena or forms of disease; indeed no disease but increase or diminution of strength would take place, on the supposition that an equal increase or diminution of excitement all over the body, were produced by the hurtful powers causing the disease.

From what has been said, it necessarily follows, that every stimulus will not be equally efficacious in curing every form of disease; which is sufficiently confirmed by experience. Hence there may be some ground for the appellation of specifics, as some medicines may act more powerfully upon the part which is the principal seat of the disease, than others do.

In the cure of diseases we ought always to attend to two things most carefully: first, to employ the proper kinds of powers, and then not to overdo them, so as to convert either diathesis into the other; and by passing over the line of health, instead of the intended cure, to substitute one disease instead of another, and thereby bring life itself into danger.

LECTURE XIII. ON THE GOUT.

There is no disease, with which the human race is afflicted, whose nature has been more mistaken than that which is to form the subject of our present consideration. It has been regarded by most practitioners as a salutary effort of the body to expel some hurtful cause, and restore health; and therefore has been looked upon as desirable to the patient. To attempt to cure it, therefore, would have been wrong, had it been curable; but it has likewise been looked upon as beyond the reach of medicine, or perfectly incurable; and, on both these accounts, after having tried a variety of drugs, without any good effect, the physicians have at last abandoned their patients, to the care of patience and flannel, which, if the constitution be not very much shattered, will often see them through the disease.

But that it is a salutary disease I deny; and I affirm, that it restores health in no other way, than the indigestion of a habitual dram drinker would be relieved by a disease in the throat, which would, for a time, prevent his swallowing any more liquor; the consequence would be, that his digestive powers would recover their tone, and he would, after a few weeks, feel himself better.

In the same way the pain and fever, which attend gout, and at the same time the inability to move, with the weakened stomach, and bad appetite, prevent the continuance of the mode of life which brought on the disease; and thus, a truce being obtained, the exhausted excitability of the body is allowed to accumulate, and the constitution, of course, feels itself renovated.

Were the disease to be viewed in this light, it is probable that many patients might in future desist from their former mode of life, which brought on the disease; and we might venture to promise them, if they did, that they would have no return of the complaint. But the misfortune is, they think the gout has restored their constitution, and that therefore they may return to their old mode of living with impunity; in consequence of which, after a few months more, the excitability is again exhausted; symptoms of indigestion come on, and the stimulant mode of living is increased, with a view to bring on the disease, which is to cure these symptoms. In this way, each time, a greater and greater degree of indirect debility is induced, and at last the system becomes so enfeebled, that the asthenic inflammation is not confined to the extremities, but attacks the head, the stomach, the lungs, and often puts a period to the existence of the patient, which has for some time been miserable.

Besides, the idea, that the gout is incurable, is a false, and a very dangerous doctrine; this is very far from being the case, and I am firmly persuaded, not only from the nature of the disease, but from experience, that it may always be cured, if taken in time, and proper directions be followed. If, by the cure of gout be meant the administration of some pill, some powder, or some potion, which shall drive away the complaint, I firmly believe, that it never was, nor ever will be cured. Indeed, it is astonishing that such an idea should have ever entered the mind of any person, who has any knowledge of nature, or particularly of the human frame; for, if the gout is a disease of indirect debility, and the effect of intemperance, as will be shown by and by, then a medicine to cure it must be something to enable a man to bear the daily effects of intemperance, during his future life, unhurt by the gout, or any other disease; that is, it must be something given now, that will take away the effects of a future cause; as well might a medicine be given to prevent a man breaking his leg, or his arm, seven years hence.

But no rational physician, or surgeon, would give a medicine with this view, in such a case as I have supposed; on the contrary, he would caution his patient against mounting precipices, scaling walls, or bringing himself again into a situation, such as produced the accident; and if he took his advice, he would, in all probability, escape a broken limb in future.

In the same way a rational physician would advise a person recovering from gout, to abstain totally and entirely from the course of life which brought it on; and this being complied with, we might venture to predict, with as much certainty in the one case as in the other, that he would in future escape it.

What I have frequently endeavoured to inculcate in the course of these lectures, always appears to me of the utmost importance: I mean, the general diffusion of physiological knowledge, or a knowledge of the human frame; this knowledge ought to form a part of general education, and is, in my opinion, as necessary for a person to learn as writing, or accounts, or any other branch of education; for if it is necessary that a young man should learn these, that he may be able to take care of his affairs, it surely can be no less necessary, that he should learn to take care of his health; for to enjoy good health, as a celebrated practical philosopher observes, is better than to command the world.

If knowledge of this kind were generally diffused, people would cease to imagine that the human constitution was so badly contrived, that a state of general health could be overset by every trifle; for instance, by a little cold; or that the recovery of it lay concealed in a few drops, or a pill. Did they better understand the nature of chronic diseases, and the causes which produce them, they could not be so unreasonable as to think, that they might live as they chose with impunity; or did they know any thing of medicine, they would soon be convinced, that though fits of pain have been relieved, and sickness cured, for a time, the reestablishment of health depends on very different powers and principles. Those who are acquainted with the nature and functions of the living body, well know, that health is not to be established by drugs; but that if it can be restored, it must be by nicely adjusting the action of the exciting powers to the state of the constitution, and the excitability; and thus gently and gradually calling forth the powers of the body to act for themselves. And though I believe that most general diseases will admit of a cure, yet I am confident, that no invalid was ever made a healthy man by the mere power of drugs. If this is a truth, should it not be universally known? If it were, there would undoubtedly be an end of quackery, for all quack medicines, from the balm of Gilead, to the botanical syrup, are supposed to cure diseases, or at least asserted to do so, in this mysterious manner.

Dr. Cullen, in his Nosology, gives us the following definition of the gout.

"Morbus haereditarius, oriens sine causa externa evidente; sed praeeunte plerumque ventriculi affectione insolita; pyrexia; dolor ad articulum, et plerumque pedis pollici, certe pedum et manuum juncturis, potissimum infestus; per intervalla revertens, et saepe cum ventriculi et internarum partium affectionibus alternans."

Now, though this definition comprises a tolerably good general character of the disease, it contains some notions, depending on the prejudice of hypothesis, which, on a careful examination, ought not, I think, to be admitted.

In the first place, I would deny, that the gout, considered as a diseased state of the system, is hereditary. This may perhaps excite some degree of surprise; and, "I had it from my father," is in the mouth of a great majority of gouty patients.

If the diseased state of the system, which occurs in gout, were hereditary, it would necessarily be transmitted from father to son; and no man, whose father had it, could possibly be free from it. There are, however, many instances to the contrary. Our parents undoubtedly give us constitutions similar to their own, and there is no doubt, that if we live in the same manner in which they did, we shall have the same diseases. This, however, by no means proves the disease to be hereditary.

We shall hereafter see, that the gout is a disease of indirect debility, brought on by a long continued use of high seasoned food and fermented liquors. There is no doubt that particular constitutions are more liable to be affected by this mode of living than others; and if my father's constitution be such, I, who probably resemble him in constitution, shall in all probability be like him, subject to the gout, provided I live in the same way; this however by no means proves the disease to be hereditary. The sons of the rich, indeed, who succeed to their fathers estate, generally succeed also to his gout, while those who are excluded from the former, are also exempted from the latter, and for very obvious reasons, unless they acquire it by their own merit.

So that though the son of a gouty parent may have a constitution predisposing to the gout; that is, more liable to be affected by causes, which produce this disease, still, if he regulate the stimuli to the state of his excitability, he will remain exempt from it.

This distinction is of much greater importance than is generally imagined; for if a person firmly believes that the gout, as a disease, is hereditary, what will be his conduct? My father had the gout, says he, therefore I must have it; well, what cannot be avoided, must be endured; let me then enjoy a short life, but a merry one: he therefore abandons himself to a luxurious mode of life, and, if the gout be the consequence, which most probably it will, he accuses his stars, and his ancestors, instead of his own misconduct.

On the contrary, if a person be convinced that he has received from his ancestors a constitution liable to be overpowered by the use of high seasoned food, and fermented liquors, and excited into gouty action, what will be his conduct? Surely, if he reason at all, it must be in this way: my father was dreadfully afflicted with the gout; I have frequently witnessed his sufferings with the deepest concern. But is not my constitution, which resembles his, liable to be affected in the same manner, by similar causes? To avoid his sufferings, therefore, I must be very temperate; more so than those who have not the hereditary propensity; for the exciting powers, which would only keep them in health, would, if applied to me, infallibly bring on the gout. In consequence of this reasoning, he adopts a temperate mode of living, and avoids the disease.

From this you must be convinced, that it is not a matter of small moment to determine, whether the gout is hereditary, and consequently unavoidable, or not. The next part of Dr. Cullen's definition is "oriens sine causa evidente". This too, I can have little hesitation to pronounce erroneous. The cause of gout, namely, the use of highly seasoned food, and the use of fermented liquors, with, in general, a luxurious, and indolent mode of living, are quite evident enough in most gouty cases, and are amply sufficient to produce the disease.

There is another part of the definition, likewise, to which I would object, as it gives a false idea of the nature of the disease, and therefore causes the preventative plan to be pursued with less confidence. I mean that part where he says "per intervalla revertens."

That the gout, when once cured, is apt to return, if the mode of life which brought it on be not abandoned, no one will deny; nay, the fits will increase in violence, because the constitution gets more and more debilitated. This, however, is not peculiar to the gout, but common to most diseases.

In describing a broken leg, it would surely be wrong to say, that it is a disease which returns at intervals, after being cured; yet, it will return as infallibly as the gout, if a person take the same kind of leap, or expose himself to the same accidents as those which brought it on. Let those, therefore, who wish to avoid a return of the gout, totally change their mode of living: otherwise, if the attacks return, let them blame themselves, and not the nature of the complaint.

These observations were thought necessary, with a view to do away some prejudices, which very much retarded our inquiries into the nature and cure of this disease. I shall now proceed to give an account of the symptoms by which it is usually attended.

The gout generally attacks the male sex; but it sometimes, though more rarely, attacks also the female, particularly those of robust and full habits. It does not generally make its appearance, till the period of greatest strength and vigour is past; for instance, about the fortieth year; but, in some cases, where the exciting causes have been powerfully applied, or where the hereditary predisposition is very strong, it attacks much earlier; such cases are, however, comparatively rare, and can, in general, be easily accounted for.

This disease is seldom known to attack persons employed in constant bodily labour, and who live temperately; and is totally unknown to those who use no wine or other fermented liquors.

If then a person of a full strong habit have for several years accustomed himself to full diet of animal food, and a regular use of wine, and malt liquor, though he may for a long time find that he can perform all the functions with vigour, his strength will at last fail: the mind and body become affected with a degree of torpor and languor for which he cannot account, and the functions of the stomach become more or less disturbed. The appetite becomes diminished, and flatulency, and other symptoms of indigestion are felt. These symptoms take place for several days, and sometimes for several weeks before the fit comes on; but often, on the day immediately preceding it, the appetite becomes greater than usual.

In this state, if the person have fatigued himself by violent exercise, or if he have exposed the extremities to cold, or if his mind have been particularly affected by any anxiety, or distressing event; or in short, if any directly debilitating cause have been applied, the fit will often follow. It sometimes comes on in the evening, but more commonly, about two or three o'clock in the morning; the pain is felt in one foot, most commonly in the ball or first joint of the great toe; but sometimes in the instep, or other parts of the foot. With the coming on of this pain there is generally more or less of a cold shivering, which as the pain increases, gradually ceases, and is succeeded by heat, which often continues as long as the pain; from the first attack the pain becomes by degrees more violent, and continues in this state, with great restlessness of the whole body, till next midnight, after which it gradually remits, and after the disease has continued for twenty four hours from the commencement of the first attack, it often ceases, and with the coming on of a gentle perspiration allows the patient to fall asleep. The patient on coming out of this sleep in the morning finds the part affected with some degree of redness and swelling, which, after having continued for some days, gradually abate.

Still however, after a fit has come on in this manner, although the violence of the pain after twenty four hours, by the excitement that it produces, cures itself, and is considerably abated, the patient is seldom entirely relieved from it. For several days he has every evening a return of considerable pain and fever, which continue with more or less violence till morning. This return is owing to the exhaustion of the excitability by the stimuli of the day, and its remission is caused by the accumulation of the excitability, by sleep.

After having continued in this manner for several days, the disease often goes off, and generally leaves the person in much better health, and enjoying greater alacrity in the functions of both body and mind, than he had for some time experienced. This is owing to the general excitement produced by the pain, which removes the great torpor and debility which preceded the fit; and from the inability to take exercise or food, the excitability accumulates again. This is the true explanation: it does not depend on any morbid matter, which the gout hunts from its lurking places, drives to a joint, and thence out of the body, as has been imagined by many.

At first the attacks of the disease are confined to one foot only: afterwards both feet become affected, though seldom at the same time; but when the inflammation appears in one, it generally disappears in the other, and as the disease continues to recur, it not only affects both feet at once, but is felt in the other joints, especially those in the upper and lower extremities, so that there is scarcely a joint in the body that is not on one occasion or other affected. After frequent attacks, the pains are commonly less violent than they were at first, the joints lose their strength and flexibility, and often become so stiff as to be deprived of all motion.

Concretions of a chalky or calcarious nature are likewise formed upon the outside of the joints. This arises from an inability of the capillary vessels, which ought to secrete the calcarious matter, and deposite it in the bones, to perform their office, from debility: hence by sympathy other vessels ta ke up the matter and deposite it in the wrong place. These concretions, though at first fluid, become at last dry, and firm: they effervesce with acids, and are totally, or in a great measure, soluble in them.

After this short description of the gout, when it occurs in its regular form, as it is called, I shall now proceed to inquire how the exciting causes produce this disease, and what is the state of the body under which it occurs.

The gout seldom occurs but in those who have for several years lived upon a full diet of animal food, often highly seasoned, and at the same time been in the habit of taking daily, or at least very constantly, a greater or less quantity of fermented licuors, either in the form of wine, or malt liquor, or both. The affection of the limb has all the appearance of an active inflammation: the part becomes swelled, hot, red, and intolerably painful. It is this circumstance which has misled practitioners, who have supposed it a case of sthenic, or active inflammation: not only the appearance, but the causes which produced it, induced them to think so; hence they were naturally led to employ the debilitating plan: a little time and observation would, however, be sufficient to convince them of its inefficacy. They would find that the application of leeches to the part, and of the lancet to the arm, instead of subduing the inflammation, would increase it: or if it did not, that the pain often attacked some internal part, which was ascribed to a translation of the morbific matter from one part to another, but which is merely owing to an increased debility: a little attentive observation would convince practitioners, however mysterious it might seem to them, that this violent inflammation was not to be cured by debilitating: on the contrary, they would see cases, in which the patient, though contrarily to the strict orders of his physicians, could not forego his old habits; but would take his wine as usual, or in greater quantity, after a few days abstinence; and this abstinence having in some degree accumulated the excitability, he would find himself much relieved by wine, and would exultingly tell them, that they were mistaken. Circumstances of this kind seem to have staggered their faith a little, but still the idea of active inflammation which they believed was visible, and almost palpable, dwelt so upon their minds, that they were but half convinced. The favourite idea of increased action of the vessels of the part had so interwoven itself with every other, that we find it never lost sight of, in the indications of cure. Hence, though bleeding is not now generally practised with the lancet, yet leeches are often applied; but the most usual plan is to consign the patient to patience and flannel; strictly forbidding wine, or fermented liquors. As an exception to this general mode, it is however observed, by some practitioners, that when the stomach is weak, and when the patient has been much accustomed to the use of strong liquors, a little animal food, and even wine, may be allowable, and even necessary.

Thus has an erroneous view of the disease been the cause of an inert practice, which wavers between the suggestions of a favourite hypothesis, and the conviction of facts.

On inquiry, however, we shall find none of the increased vigour in the system, which has been suspected, nor increased action in the part more particularly affected; on the contrary, the whole body is in a state of indirect debility, or exhausted excitability, and the part more particularly affected, in a state of asthenic inflammation.

If the gout were of a sthenic or inflammatory nature, might we not ask, why the causes which produce it, do not produce it in the meridian of life, when they produce their greatest effect, and when real sthenic diseases are most apt to occur? or, why the symptoms of the inflammation, like all other real sthenic inflammations, are not relieved by the debilitating plan? The contrary, however, points out to us clearly the nature of the disease: the gout is not a sthenic disease, or a disease of strength: it does not depend upon increased vigour of the constitution, and plethora, but is manifestly asthenic, like all the rest of the asthenic diseases. The mode of living is such as brings on indirect debility, or exhaustion of the excitability, such as the use of rich and highly seasoned food, and a daily use of fermented liquors. These at first certainly produce vigour, or strength, and will be the cause of sthenic diseases; but they are generally taken in such a manner, that, though they produce a degree of excitement above the point of health, still they only approach the line of sthenic disease, without in general falling into it. They continue, however, to exhaust the excitability, and by the time that the vigour of the body begins naturally to decline, the system of a person who has lived in this manner is unusually torpid; all the blood vessels, which have hitherto been distended with rich blood, begin to lose their tone, from their excitability having been exhausted by the use of these powerful stimulants; but this torpor is particularly and first experienced in those parts which have been more immediately subject to the action of the exciting causes; viz. the stomach and bowels: symptoms of indigestion occur, and the excitability of these organs having been almost entirely exhausted by the violent action of the stimulants applied, cannot now be roused to any healthy action; the food is not properly digested, but runs into a kind of fermentation, which causes an extrication of gas: this distends the stomach and bowels, and produces pains, uneasy eructations, and all the distressing symptoms of indigestion. Nor is this in the least surprising, when we consider that many people who have brought on complaints of this kind, have been in the habit of eating heartily of rich and highly seasoned animal food, and of drinking from a pint to a bottle of wine, and perhaps a quantity of malt liquor, almost every day of their lives for years. This mode is sufficient to wear out the powers of the stomach, were it three times as capacious as it is, and of the constitution, were it ten times as strong.

When a torpor, or state of exhausted excitability, of the whole system, has been induced in this manner, and symptoms of indigestion produced, any directly debilitating cause applied to the extremities, adding to the indirect debility, causes a total torpor, or inactivity of the minute vessels of the part, and thus totally destroys the balance between the propelling and resisting force; hence the vessels will be morbidly distended with blood, a swelling and redness will take place, and an asthenic inflammation, produced in the way which I fully pointed out in the last lecture, will be established. Hence the pain, and other symptoms, which accompany a fit of the gout. Hence likewise we see, why debilitating powers applied to the part will not reduce the inflammation; and why a warmth, which aggravates every really sthenic inflammatory affection, is so comfortable in this.

Almost any debilitating cause, when the system has been brought by intemperance to the torpid state, which I have described, will bring on a fit of the gout, but nothing more certainly than cold or moisture: hence if a person have his feet chilled or wet, he will be almost certain to have an attack.

Hence we see that the asthenic inflammation is not the disease, but merely a symptom of it; and like other symptoms, fallacious in its appearance; the disease is a state of indirect debility, to which our attention ought to be directed.

When this inflammation is violent, and accompanied with great pain, after several hours continuance, it excites the action of the minute vessels, enables them to propel the blood, by which they are morbidly distended, and restores the balance between the resisting and the propelling force; and thus the inflammatory appearances will for a time subside, but the torpor of the whole system remaining, and the debility of the vessels returning, when their excitement, which was the consequence of their action, has ceased, another asthenic inflammation will take place, which will again cure itself as before; so that during a paroxysm, several remissions will take place, as was mentioned in the description of the disease. As, during the paroxysm, the pain causes a considerable degree of excitement over the whole system, the action of the stomach and other parts is roused by it; during the fit likewise, little nutriment is taken, so that by the action of the stomach and bowels, they get rid of their load; rest likewise assists to accumulate the excitability, so that from all these causes together, the body becomes restored to a state of vigour, which, compared with its former torpidity, makes the patient imagine that this friendly disease has restored him to a state of unusual health, and even renovated the powers of his constitution. Under this mistaken idea, he does not, when the fit leaves him, abandon the mode of life, which brought on the disease; highly seasoned food, and the usual quantity of wine, are again resorted to: after a time the torpor of the system, and symptoms of indigestion return, and he again hopes that his friend the gout will come and cure him.

By a continuance of this plan, the inflammation again appears; but the system having become more torpid, the inflammatory action is by no means so great as it was before: if it has power to restore the equilibrium between the resistance and propelling force, and thus cure itself, this effect is entirely confined to the inflamed part.

The other foot labouring under similar torpor, or debility, now feels the effects of the propelling force, and an inflammation takes place in it, which having cured itself in the same manner, and the torpor of the foot first affected being returned, or even greater than it was before, on account of the previous excitement; the inflammation again attacks this foot, and thus the gout is supposed to emigrate from one limb to another. The gout, as a disease of general debility, however, remains the same; and it is only these symptoms, which form but a small part of the disease, that vary according to circumstances.

If, during an asthenic inflammation of the lower, or upper extremities, the torpor and debility of the whole system increase, then the force of the circulation, or propelling force, being diminished, the symptoms of inflammation will suddenly disappear; but as great debility now prevails, the stomach will be apt to be affected with cramps or convulsions, or an asthenic inflammation of some internal part will take place: for, though the propelling force is not sufficient to overdistend the debilitated vessels of the extremities, it will distend those of the internal parts nearer the heart, which are now debilitated.

In this case, it has been generally, but absurdly imagined, that the gout is translated, or recedes from the extremities to some internal part: the term of retrocedent gout has therefore been applied to occurrences of this nature. From the explanation which has been given, it is evident, that this term is improper. The general debility being increased, the propelling force becomes unable to produce an inflammation of the extremities, and this is the reason why it disappears. The disease, however, is not at all altered in its nature by this variation of symptoms. It is still the same, by whatever name it may be called.

It sometimes happens, that after full living, the stomach becomes particularly affected, and the patient is troubled with flatulency, indigestion, loss of appetite, eructations, nausea, and vomiting, with great dejection of spirits, pain and giddiness of the head, disturbed recollection, or muddiness of intellect, as it is termed, with all the symptoms, which usually precede a regular fit of the gout, yet no inflammatory affection of the joints is produced. This state has been absurdly enough called the atonic gout, as if there were a gout accompanied with vigour and sthenic diathesis: but the absence of inflammation in the extremities may depend on two causes. First, the powers producing the disease, may have debilitated the stomach and first passages, while the vessels of the extremities are not particularly debilitated, and the resisting force is able to counterbalance the propelling force: in this case, no morbid degree of distention or inflammation of the extreme vessels can take place. Secondly, the general debility may be such, and the power of the circulation so much diminished, that, though the extreme vessels may be debilitated, no inflammation, or preternatural distention will take place.

Hence, we see, that this is still the same disease; but that physicians have erred in their explanation of the symptoms, by regarding that as the principal part of the disease, which is only a symptom.

We have seen then, that by the theory which has been unfolded, all the symptoms of this hitherto mysterious disease are plainly and naturally explained. We shall next see if the only method of cure which experience warrants, cannot be explained upon the same principles.

If, on entering this part of the subject, any one should expect that I should furnish him with a receipt, consisting of certain drugs, which swallowed, will cause this terrible disease to disappear, and health to take its place, he would be very much mistaken; for, can any person in his senses suppose that a disease, which he has been almost his whole life in contracting, and an exhausted state of the excitability, which has been gradually brought on by years of intemperance, can be dispersed by a pill, a powder, or a julep? Or, if the symptoms could be relieved by medicine, which they often may, can he suppose, that they will not return, if the same mode of living, which first brought them on, be continued?

I shall, however, proceed to give some directions, which if rigidly persevered in, will not only afford relief in the fit, but will prevent its return with such violence, and at last totally eradicate it, provided the constitution be not completely exhausted, and almost every joint stiffened with calcarious concretions.

The inflammation of the extremities may at any time be relieved by means of electricity, or by stimulant embrocations applied to the part, and this without any danger whatever of throwing the complaint on some more vital part, as has generally been imagined. If I were to apply any debilitating means to the part, I should then probably relieve the pain; but, by debilitating the whole system, should cause an attack of the stomach, or some other internal part, as has been already explained; but by a stimulant application to the inflamed part I run no such risk. The inflammation is of the asthenic kind, depending upon a debility of the small vessels, whereby they do not afford sufficient resistance to the propelling force, and therefore become morbidly distended, or inflamed, as it is termed, though this term is certainly improper, even in a metaphorical view: but a stimulant application to the part excites the debilitated vessels to action; their contraction diminishes the morbid quantity of blood; and the balance between the propelling and resisting forces being restored, the inflammation of course ceases. This is not a mere deduction, a priori, from the theory of inflammation, which I have delivered; it is the result of repeated experience. I have seen several very violent gouty inflammations very speedily removed by electricity. Small sparks should be drawn from the part affected, at first through flannel, and increased as the patient can bear them: sparks alone are necessary; recourse need never be had to shocks. But though we thus remove a very painful part of the disease, yet still a formidable debility remains, and unless this be removed, the inflammation will be apt to return. In endeavouring to remove this general debility, we must recollect, that it is of the indirect kind, or depends upon an exhausted state of the excitability; our great object therefore, is to allow the excitability to accumulate. But this accumulation depends as well upon the proper action of the different functions, as upon the withdrawing of stimulants: we ought therefore to guard carefully against costiveness, by which the proper action of the stomach and bowels is very much injured: but we must use warm laxatives. An infusion of senna and rhubarb in proof spirits, made still stronger by aromatics, has always seemed to me to answer the purpose best, and this should be taken of a temperature rather above blood warm; for instance, about 100 degrees. This is particularly necessary, when the gout attacks the stomach, and I have several times seen a severe attack of it removed in half an hour, by a tincture of this kind. Indeed, the most violent attacks of the stomach may be relieved; and are only to be relieved by spirits, ether, and opium.

It is on this organ, that the hurtful powers have produced their greatest effect; for to it they are immediately applied. It is by no means surprising, that the constant application of highly seasoned foods, with fermented and spirituous liquors, should at last wear out the vital principle of this organ. Indeed it is often so far exhausted, that the most terrible cramps and convulsions take place, which would soon end in its total extinction, unless it were roused to somewhat like a proper action by the most powerful stimulants. Still, however, their effect is but temporary.

With respect to a regular fit, after the inflammation of the extremities has been subdued by the means I have mentioned, a generous, but not full diet should be used. A person who has been for a long time accustomed to wine, cannot easily be deprived of it at once; but he should drink Madeira, and those wines, which neither contain much carbonic acid, nor deposite much tartar. His food should be of the plainest kind, and generally boiled, instead of roast. The great thing is to keep the spirits and excitement rather under par, but not to let the patient sink too low. In this way, the exhausted excitability will gradually accumulate, and the healthy state be reestablished. When this is once effected, the gout may be prevented in future with the greatest certainty, if the patient will have resolution. The whole secret consists in abstaining, in toto, from alcohol, in every form, however disguised, or however diluted. He must not take it, either in the form of liqueurs, cordials, wine, or even small beer.

I believe there never was an instance of a person having the gout, who totally abstained from every form of alcohol, however he might live in other respects: and I doubt very much, if ever the gout returned after a person had abstained from fermented or spirituous liquors for two years.

Temperance in eating, and exercise, are, no doubt, powerful auxiliaries, and tend very much to promote health; but still they will not secure a person from a return of the gout, without this precaution. There seems something in alcohol, which peculiarly brings on this state of the constitution, and without it, it would seem that gout could not be produced. Here then is an effectual method of curing the gout, which will no more return, if this method be strictly persevered in, than the smallpox will attack the constitution after inoculation.

During the fit therefore, I would say, nearly in the words of Dr. Darwin, Drink no malt liquor on any account. Let the beverage at dinner consist of two glasses of Madeira, diluted with three half pints of water; on no account whatever drink any more wine or spirituous liquors in the course of the day. Eat meat constantly at dinner, without any seasoning, but with any kind of tender vegetables, that are found to agree. When the fit is removed, use the warm bath twice a week, an hour before going to bed, at about 93 degrees, or 94 degrees of heat. Keep the body open by means of lenitive electuary and rhubarb; for there is an objection to the tincture I mentioned, as containing alcohol. Use constant, gentle exercise; but never so violent as to bring on great fatigue. The grand secret, however, in the cure, as has been already observed, but which cannot be too often inculcated, is to abstain, in toto, from every thing that contains alcohol.

In short, though in acute diseases medicines are highly useful, a chronic disease can never be cured, and the healthy state reestablished, by them alone. To effect a cure in such cases, we must reform our mode of life, change our bad habits into good ones; and then, if we have patience to wait the slow operations of nature, we shall have no reason to regret our former luxuries.

LECTURE XIV.
NERVOUS COMPLAINTS, &c.

In this lecture I propose to take a view of some of those affections, which have been commonly, but improperly known by the appellation of nervous complaints, because it has been supposed by many that they are owing to a deranged state of the nerves, which, however, is by no means the case; for I hope to be able to make it appear, that these symptoms arise from a general affection of the excitement of the system. In short, by far the greater number of these complaints, arise from such a state of the excitement as approaches predisposition, or perhaps ranges between predisposition and disease, but does not in general actually reach disease; or rather, it is a state of the excitement, so far departing from the point of perfect health, that the functions are not performed with that alacrity, or vigour, which ought to take place; but labour under that disturbed and uneasy action, which, though it cannot be called actual disease, yet deviates considerably from the point of perfect health.

This is a new view of these diseases, but the more I have examined it, the more I am convinced that it is just. Indeed, the name, nervous, has generally been given to an assemblage of symptoms, which the physicians did not understand; and when the patient relates a history of symptoms, and expects that his physician shall inform him of the name, and nature of his complaint, he generally receives for answer, that his complaints are nervous, or bilious; terms which convey no distinct ideas, but which serve to satisfy the patient, and to conceal the ignorance of the physician, or spare him the labour of thinking.

Indeed, the idea of nervous diseases, which I have already pointed at, is not only new, but could only have arisen from such a view as we have been taking of the states of excitement and excitability. This view will not only lead us to form a more just idea of the manner in which these diseases originate, but will point out a distinction of them into two classes, of the utmost use in practice, but which distinction has totally escaped the attention of practitioners; for though these complaints have been generally thought to arise from a lowness of the nervous energy, or some kind of debility, or weakness of the nervous system, and, on this account, the stimulant and cordial plan of cure has been recommended, I am convinced, from observation, that nearly one half of them, if not more, originate from a state of the excitement verging towards sthenic disease; and in these cases, this general mode of treatment must be highly improper.

It has been already shown, that when the common exciting powers which support life, act in such a manner, that a middle degree of exciting power, acts upon a middle degree of excitability, the most perfect state of the system, or a state of perfect health, takes place: it is, however, seldom in our power so to proportion the state of excitement and excitability to each other. The action of the exciting powers is continually varying in strength; and the excitability, from a variety of stimulants, and other circumstances, which are not entirely under our direction, is sometimes more, and sometimes less abundant, than this middle degree. There is, however, a considerable latitude, on each side the point of health, within which the excitement may vary, and yet no disease, nor any disturbance of the functions may take place: but this has its limits, beyond which if the excitement be brought, on either side, it is evident that an uneasy or unpleasant exercise of the functions must take place. There is not, however, any precise line or boundary between this state, and that in which the functions begin to be disturbed; or the contrary, the law of continuity and gradation seems to extend throughout every part of nature. This departure from the healthy state, and approach to disease, in which what has been called the nervous state consists, is gradual and scarcely perceptible; but is apt to be produced by any circumstances, which lead the excitement beyond its proper limits.

Nervous complaints may therefore be divided, like all other diseases, into two classes. First, those in which the excitement is increased, or in which it verges to, or has actually reached, the point of predisposition to sthenic disease; Secondly, those in which the excitement is diminished, or in which it verges towards asthenic disease. This last class, as has been done before, may be subdivided into two orders. The first will comprise those diseases in which the excitability is sufficiently abundant, or even accumulated, but where the excitement is deficient from a want of energy in the exciting powers. In the second, there has been no deficiency in the action of the exciting powers; but on the contrary, probably for a considerable time, some of the diffusible stimuli not natural to life have been applied; in this case, the excitability has become exhausted, and a proper degree of excitement cannot be produced by the action of the common exciting powers.

No diseases show so clearly the fallacy of trusting to symptoms, as those of the former class. I have met with innumerable cases of this kind, in which, if you were to trust to the patients own description, they laboured under considerable debility; and had it not been for the particular attention I paid to my own case, I should not probably have suspected that a directly opposite state of the system may produce these symptoms.

From inheriting a good constitution, and being brought up in the country in a hardy manner, I am so much predisposed to the sthenic state, that I may consider the state of my excitement, as generally, indeed almost always, above the point of health: and unless I live in the most temperate, and even abstemious manner, the excitement is extremely liable to overstep the bounds of predisposition, and fall into sthenic disease. I have had several attacks of this kind of disease; and indeed, I never remember to have laboured under any disease of debility, or diminished excitement.

Health, according to the view we have taken of it, may be compared to a musical string, tuned to a certain pitch, or note; and though perhaps in the great bulk of mankind, either from the manner of living, or from other circumstances, the excitement is a little below, and requires to be screwed up to the healthy pitch, yet there are others where it is apt to get constantly above, and where it requires letting down to this pitch; my constitution is one of these: but I have this consolation, that if I can for a few years ward off the fatal effects of some acute sthenic diseases, this tendency to sthenic diathesis will gradually wear off, and I may probably enjoy a state of good health, at a time, when most constitutions of an opposite cast begin to give way. Whenever I have for some time lived rather fully, though by no means intemperately, after having for some days, or perhaps some weeks experienced an unusually good flow of spirits, and taken exercise with pleasure, I begin, first of all, to have disturbed sleep, I find myself inclined to sleep in the morning, as if I had not been refreshed by the night's sleep; my spirits become low, and I am apt to look upon the gloomy side of every thing I undertake or do. I feel a general sense of languor and debility, and am ready, as I have heard many patients labouring under the same state exclaim, to sink into the earth. From the slightest causes, I am apt to apprehend the most serious evils, and my temper becomes irritable, and scarcely to be pleased with any thing. If in this state, I take exercise, I soon feel myself fatigued; a disagreeable stupor comes on, without, however, the least degree of perspiration, and I feel an inability to move.

At first, I used to imagine these to be symptoms of debility, or diminished excitement, nor was it till after several ineffectual trials to relieve them by the tonic, or stimulant plan, that I was convinced of my mistake. This plan always caused an aggravation of every symptom, and if I persevered in it, an inflammatory disease was sure to be the consequence. Indeed, I might have suspected this, from considering, that these symptoms had been brought on by full living, and preceded by good spirits; but my mind had received such a prejudice from the writings of medical men, who had uniformly described these as a train of nervous symptoms, as they called them, depending on a debilitated state of the nervous system, that I was blind to conviction, till repeated disappointment from the stimulating plan, convinced me I must be wrong. The only alternative therefore, was a contrary plan, and the immediate relief I experienced, was a proof that I had detected the real nature of the complaint. Since that time, I can at any time prevent these unpleasant symptoms, by an abstemious course of life, and remove them, when they have come on, by the debilitating plan; which, instead of weakening, gives additional elasticity and strength to the fibres, and alacrity to the spirits. I have described the symptoms in one case, as this will serve as a general description. We may add, that persons labouring under this kind of predisposition, are particularly attentive to the state of their own health, and to every change of feeling in their bodies; and from every uneasy sensation, perhaps of the slightest kind, they apprehend great danger, and even death itself. In cases of this kind, the bowels are generally costive, and the spirits of the patient are very apt to be affected by changes in the weather, particularly by a fall of the barometer. How the diminution of atmospheric pressure acts in increasing the symptoms, we perhaps do not know; but its effects are experienced almost universally.

It is evident, that the only mode of cure in cases of this kind is extreme temperance: animal food should be taken sparingly, and wine and spirits in general totally abstained from. The bowels should be kept open by any mild neutral salt. I have generally found magnesia and lemonade to agree remarkably well in such cases. Exercise on horseback, is also particularly useful; bark, bitters, and the fetid and antispasmodic medicines, which are generally prescribed in such cases, are extremely hurtful.

This view of nervous complaints is, I may venture to say, as new as it is just. It has never been imagined, that any of them depended upon too great excitement; on the contrary, they have been universally considered as originating in debility, and of course, tonics were prescribed, which, though they produced the greatest benefit in the other class of nervous complaints, in these they occasioned the most serious evils, and often brought on real inflammatory diseases, or even diseases of indirect debility, as I have repeatedly seen.

These cases cannot at first sight, however, be easily distinguished from those of the opposite class; the symptoms being nearly alike, and the patient complaining of languor, debility, and extreme depression of spirits in both. But by attending carefully to the effects produced by the exciting powers, they may in general be distinguished. A patient of this kind will tell you, that he does not feel pleasant effects from wine, or spirituous liquors; instead of exhilaration, his spirits become depressed by them; whereas, in the contrary state, he finds almost instant relief. By attending to circumstances of this kind, the nature of the complaint may in general be ascertained.

Highly seasoned, and strongly stimulant foods should in the sthenic hypochondriasis, if it may be so called, be avoided; but the most mischievous agent of all, and which contributes to bring on the greater number of these complaints, is wine. This, I believe, produces more diseases, than all other causes put together. Every person is ready to allow, that wine taken to excess is hurtful, because he sees immediate evils follow; but the distant effects, which require more attentive observation to perceive, very few see, and believe; and, judging from pleasant and agreeable feelings, they say that a little wine is wholesome, and good for every one; and accordingly take it every day, and even give it to their children; thus debauching their natural taste in the earliest infancy, and teaching them to relish what will injure their constitutions; but which, if properly abstained from, would prove one of the most valuable cordial medicines we possess.

The idea that wine or spirituous liquors assist digestion, is false. Those who are acquainted with chemistry, know that food is hardened, and rendered less digestible by these means; and the stimulus, which wine gives to the stomach, is not necessary, excepting to those who have exhausted the excitability of that organ, by the excessive use of strong liquors. In these, the stomach can scarcely be excited to action, without the assistance of such a stimulus. If the food wants diluting, water is the best diluent. Water is the only liquor that nature knows, or has provided for animals; and whatever nature gives us, is, we may depend upon it, the best, and safest for us. Wine ought to be reserved as a cordial in sickness, and in old age; and a most salutary remedy would it prove, did we not exhaust its power by daily use.

I am sensible that I am treading on delicate ground, but I am determined to speak my sentiments with plainness and sincerity, since the health and welfare of thousands are concerned. Most persons have so indulged themselves in this pernicious habit of drinking wine, that they imagine they cannot live without a little every day; they think that their very existence depends upon it, and that their stomachs require it. Similar arguments may be brought in favour of every other bad habit. Though, at first, the violence we do to nature makes her revolt; in a little time she submits, and is not only reconciled, but grows fond of the habit; and we think it necessary to our existence: neither the flavour of wine, of opium, of snuff, or of tobacco, are naturally agreeable to us: on the contrary, they are highly unpleasant at first; but by the force of habit they become pleasant.

It is, however, the business of rational beings to distinguish carefully, between the real wants of nature, and the artificial calls of habit; and when we find that the last begin to injure us, we ought to use the most persevering efforts to break the enchantment of bad customs; and though it cost us some uneasy sensations at first, we must learn to bear them patiently; a little time will reward us for our forbearance, by a reestablishment of health and spirits.

I shall now proceed to examine the opposite class of nervous complaints: or such as do really depend on debility, or an asthenic state of the system. These may be divided into two orders; viz. those of direct, and those of indirect debility. I shall first consider those of direct debility.

Though these complaints originate from a deficiency of stimulus, yet it is very seldom from a deficiency of the common stimulant powers. The only people, who in general labour under this deficiency of the common stimulants, are the poor; they are seldom troubled with nervous complaints; their daily exercise, and constant attention to procure common necessaries, prevent their feeling what so grievously afflicts the rich and luxurious. These complaints arise chiefly from a deficiency of mental stimuli. The most common cause of them, and whose effects are the most difficult to remove, is to be looked for in the mind.

The passions and emotions, when exercised with moderation, and kept within proper bounds, are the sources of life and activity; without these precious affections, we should be reduced to a kind of vegetation, equally removed from pleasure and from pain. For want of these elastic springs, the animal spirits lose their regularity and play; life becomes a lethargic sleep, and we fall into indifference and languor.

If then the passions are so necessary to the support of the health of the body, when in a proper degree, can we expect, that when they are inordinate or excessive, or even deficient, we shall escape with impunity? tumultuous passions are like torrents, which overflow their bounds, and tear up every thing before them; and mournful experience convinces us, that these effects of the mind are easily communicated to the body. We ought, therefore, to be particularly on our guard against their seduction.

"'Tis the great art of life to manage well
The restless mind."

It is particularly in their infancy, if it may be so called, that we ought to be upon our guard against their seduction; they are then soothing and insidious; but if we suffer them to gain strength, and establish their empire, reason, obscured and overcome, rests in a shameful dependence upon the senses; her light becomes too faint to be seen, and her voice too feeble to be heard; and the soul, hurried on by an impulse to which no obstacle is presented, communicates to the body its languor and debility. The passions, by which the body is chiefly affected, are, joy, grief, hope, fear, love, hatred, and anger. Any others may be reduced to some of these, or are compounded of them. The pleasurable passions produce strong excitement of the body, while the depressing passions diminish the excitement; indeed it would seem that grief is only a diminution of joy, as cold is of heat; when this passion exists in a proper degree, then we feel no particular exhilarating sensation, but our spirits and health are good: we cannot doubt, however, that we are excited by a pleasant sensation, though we are unable to perceive it. In the same manner, when heat acts moderately, or is about the degree we call temperate, we do not perceive its effects on the body, though there can be no doubt, that the body is under the influence of its stimulus, and powerfully excited by it; for when it is diminished, or cold applied, we feel a deficiency of excitement, and become afterwards more sensible of heat afterwards applied.

The same takes place with respect to joy and grief, and proves, I think, clearly, that the one is only a diminution of the other, and that they are not different passions. When the body has been exposed to severe cold, the excitability becomes so much accumulated with respect to heat, that if it be afterwards applied too powerfully, a violent action, with a rapid exhaustion of the excitability, which ends in mortification, or death of the part, will take place. We should therefore apply heat in the gentlest manner possible, and gradually exhaust the morbidly accumulated excitability.

In the same manner, when the body has been under the influence of violent grief, any sudden joy has been known to overpower the system, and even produce instant death. We have an instance in history, of a mother being plunged into the extreme of grief, on being informed that her son was slain in battle; but when news was brought her, that he was alive, and well, the effect upon her spirits was such, as to bring on instant death. This event ought to have been unfolded to her in the most gradual manner; she should have been told, for instance, that he was severely wounded; but that it was not certain he was dead; then that there was a report he was living, which should have been gradually confirmed, as she could bear it. The same observations may be made, with respect to hope and fear, or despair; the former is an exciting passion, the latter, a depressing one; but the one is only a lower degree of the other; for a moderate degree of hope produces a pleasant state of serenity of the mind, and contributes to the health of the body; but a diminution of it weakens; and a great degree of despair so accumulates the excitability of the system, as to render it liable to be overpowered by any sudden hope or joy afterwards applied. What proves that joy and hope act by stimulating, and grief and despair by withdrawing stimulant action from the body, is, that the former exhaust excitability, while the latter accumulate it. Joy, for instance, does not render the system more liable to be affected by hope, but the reverse; and the same may be said of hope. In the same way, heat does not render the body more liable to be affected by food, but the reverse. Both these are stimulants, and exhaust the excitability. But after heat has been applied, if it be followed by cold, a great degree of languor or weakness will take place; because we have here a direct debility, added to indirect debility. In the same way, grief succeeding joy, or despair succeeding hope, produce a greater degree of dejection, both of mind and body, than if they had not been preceded by these stimulant passions; because here, direct debility is added to indirect. The excitability is first exhausted, and then the stimulus is withdrawn.

We see then, that the passions of the mind act as stimulants to the body, that, when in a proper degree, they tend to preserve it in health; but when their action is either too powerful, or too small, they produce the same effects as the other powers. We should therefore naturally expect, that when there is a deficient action of this kind of mental stimulus, or when the mind is under the influence of the depressing passions, a predisposition to diseases of direct debility would take place, and even these diseases be produced. Accordingly we find a numerous class of nervous complaints originating from these causes. Indeed, the undue action of the mental stimulants, produces more quick alterations in the state of the excitement, than that of the other exciting powers. Violent grief, or vexation, will immediately suspend the powers of the stomach. If we suppose a person in the best health, and highest good humour, sitting down to dinner with his friends, if he suddenly receives any afflicting news, his appetite is instantly gone, he cannot swallow a morsel. If the same thing happens after he has made a hearty dinner, the action of the stomach is suddenly suspended, and the whole process of digestion stopped, and what he has eaten, lies a most oppressive load. But this is not all: the whole circulation of the blood becomes disturbed; the contraction and dilatation of the heart become irregular; it flutters, and palpitates; hence all the secretions become irregular, some of the glands acting too powerfully, others not at all; hence the increased action of the kidneys, and hence a burst of tears; hysterical affections, epilepsy, and syncope, frequently succeed, in which every muscle of the body becomes convulsed. Indeed, many terrible diseases originate from this source, which were formerly ascribed to witchcraft, and the possession of devils.

In slower, more silent, but longer continued grief, the effects are similar, but not so violent. The functions of the stomach are more gently disturbed, its juices vitiated; and acidity, and other symptoms of indigestion, will show themselves. Hence no bland and nutritive chyle is conveyed into the blood; whence emaciation and general debility must follow; and the patient will at last die, as it is said, of a broken heart.

Besides the disturbed state of the stomach, and bad digestion, there can be no sleep in this state of mind; for,

"Sleep, like the world, his ready visit pays,
Where fortune smiles; the wretched he forsakes;
Swift on his downy pinion flies from woe,
And lights on lids unsullied with a tear."

Hence the animal spirits will not be recruited, nor the worn out organs restored to vigour.

The minds of patients labouring under this division of nervous diseases, are likewise in general filled with over anxiety concerning their health; attentive to every feeling, they find, in trifles light as air, strong confirmations of their apprehensions.

It is evident, that in these cases, a state of direct debility prevails, attended with a morbidly accumulated excitability; hence, those remedies afford relief, which produce a quick exhaustion of this principle, and thus blunt the feelings, and lull the mind into some degree of forgetfulness of its woes. Hence opium, tobacco, and the fetid gums are often resorted to; and in the hands of a judicious practitioner, they will afford great relief, provided he carefully watch the patient, and prevent their abuse; for, if left to the discretion of the patient, he finds that kind of relief which he has long wished for; his moderation knows no bounds, and he is apt to take them in such a manner, as to add indirect debility, to direct, and thus bring on a state of exhausted excitability, while there is still a diminished state of mental stimulants. This will cause his spirits to be more depressed than ever; he will therefore increase the dose, whether it be of opium, tobacco, or spirituous liquors, and thus he will be hurried on, adding fuel to the flame, till his exhausted excitability becomes irrecoverable, and he ends his days in a miserable state of imbecility, if not by suicide. Hence, though some of these narcotic stimulants, which exhaust the excitability, and blunt the feelings, may be employed with advantage, in order to prepare the mind for those changes, which the physician wishes to produce, they should be used with the greatest caution, and never left in any degree to the discretion of the patient. The cure, however, depends chiefly on regulating the state of the mind, or interrupting the attention of the patient; and diverting it, if possible, to other objects than his own feeling.

Whatever aversion to application of any kind we may meet with in patients of this class, we may be assured that nothing is more pernicious to them than absolute idleness, or a vacancy from all earnest pursuit.

The occupations of business suitable to their circumstances, and situations in life, if neither attended with emotion, anxiety, nor fatigue, are always to be advised to such patients; but occupations which are objects of anxiety, and more particularly such as are exposed to accidental interruptions, disappointments, and failures, are very improper for patients of this class.

To such patients exercise in the open air is of the utmost consequence. Of all the various methods of preserving health and preventing diseases, which nature has suggested, there is none more efficacious than exercise. It puts the fluids all in motion, strengthens the solids, promotes digestion, and perspiration, and occasions the decomposition of a larger quantity of air in the lungs, and thus not only more heat, but more vital energy is supplied to the body; and of all the various modes of exercise, none conduces so much to the health of the body, as riding on horseback: it is not attended with the fatigue of walking, and the free air is more enjoyed in this way, than by any other mode of exercise. The system of the vena portarum, which collects the blood from the abdominal viscera, and circulates it through the liver, is likewise rendered more active, by this kind of exercise, than by any other, and thus a torpid state, not only of the bowels, but of this system of vessels, and the biliary system, is prevented.

When a patient of this class, however, goes out for the sake of exercise only, it does not in general produce so good an effect, as might be expected; for he is continually brooding over the state of his health: there is no new object to arrest his attention, and he is constantly reminded of the cause of his riding. Exercise will therefore be most effectual when employed in the pursuit of a journey, where a succession of pleasant scenes are likely to present themselves, and new objects arise, which call forth his attention. A journey likewise withdraws the patient from many objects of uneasiness and care, which might present themselves at home.

With respect to medicines, costiveness, which often attends these diseases, ought to be carefully avoided, by some mild laxative. Calcined magnesia, and lemonade, have always seemed to me to answer the purpose; but the most effectual method is to acquire a regular habit, which may be done by perseverance, and strict attention.

Chalybeate waters have been frequently tried, and may in general be recommended with success, particularly, as the amusement and exercise generally accompanying the use of these waters, aid the tonic powers of the iron. The bark may likewise be exhibited with advantage.

There is yet another class of nervous diseases which we have to notice, which are by no means uncommon; yet they have, like the first class, escaped the attention of writers on this subject, and of medical practitioners in general: I mean those where the system is in a state of torpor, or exhausted excitability.

This state of the system may be brought on by various causes, but principally by the long continued use of opium, tobacco, or fermented liquors.

When these substances, which are powerful stimulants, have been taken for some time, they bring on a state of the system so torpid, that the usual exciting powers, and the usual occurrences, which in general produce pleasant sensations, do not occasion a sufficient degree of excitement, in those whose excitability is thus exhausted. They therefore feel continual languor and listlessness, unless when under the influence of the stimulus which brought on the exhaustion. Every scene, however beautiful, is beheld with indifference by such patients, and the degree of ennui they feel is insupportable: this makes them have recourse to the stimulus which has exhausted their excitability, which in some degree removes this languor for a time; but it returns with redoubled strength, and redoubled horror, when the stimulant effect is over: and as this repetition exhausts the excitability more and more, the stimulus is repeated in greater quantity, and thus the disease increases to a most alarming degree.

There is no way of curing this state of nervous torpor, but by leaving off the stimuli which caused the exhaustion; and if the patient have resolution to do this for a few weeks, though, at first, he will, no doubt, find his spirits a little depressed, he will ultimately overcome the habit, and will be rewarded by alacrity of spirits, such as he never experiences under the most powerful action of artificial stimulants.

I must not, however, forget to notice, that there is a nervous state, or ennui, originating from a wrong direction of mental exertion, which exhausts the excitability to a great degree, and brings on a state of depression scarcely to be born.

When a person has by habit made his mind constantly dependent on dissipation, on gaming, and on frivolous, but not inactive pursuits, in order to produce pleasurable sensations, and at the same time neglected that culture of the understanding which will enable him to retire into himself with pleasure, and receive more enjoyment from the exercise of this cultivated understanding than he does in the most noisy, or fashionable circle of dissipation: I say, when there is this vacancy of mind, whenever it is not engaged in such pursuits as I have mentioned, a languor and weariness is experienced, which is intolerable, and which prompts the person so circumstanced, to fly continually to the only scenes which interest his mind. Hence, the passion for gaming, in which the anxiety attending it causes an interest in the mind, which takes off the dreadful languor experienced, when it is not thus employed.

It is owing to wealth, admitting of indolence, and yielding to the pursuit of transitory and unsatisfying amusements, or to that of exhausting pleasures only, that the present times exhibit to us so many instances of persons suffering under this state: it is a state totally unknown to the poor, who labour for their daily bread, and to those whose minds are actively employed in study or business. It can only be cured by cultivating the understanding, and applying to some art or science, which will engage and interest the attention. I have received the thanks of many for recommending the study of philosophy, and particularly of chemistry, to their attention. This affords a rational and interesting pursuit, which, if entered into with ardour, and if the person actually works, or makes experiments himself, he will soon experience an enjoyment and an interest, such as he never experienced at the gaming table, or at any other place of fashionable amusement. Nay, I will venture to say, that all elegant amusements will be enjoyed with much greater relish by one who employs himself in some rational pursuit, and only resorts to such amusements as a relaxation, than by one who makes these amusements a business.

From the view we have taken of these complaints, it is evident, that they are like other general diseases of the sthenic, or asthenic kind; they seem to constitute a state of the body between predisposition and disease; and they differ from most diseases in this, that in most complaints the increase, or diminution of the excitement is unequal in different parts of the body, and this gives rise to the different forms of disease; but in nervous complaints the excitement seems much more equably affected in different parts. These complaints, as we have seen, may be divided into three classes; sthenic; those of accumulated excitability; and those of exhausted excitability: but though they are evidently distinguishable in this manner, and require different modes of cure, I have never seen any account of more than one kind in any medical writer: the same remedies were prescribed for all, however different they might be.

Though medicines may relieve complaints of this kind, and particularly those of the second class, yet from what has been said, it must be evident, that much more may be done by regulating the action of the common exciting powers. Indeed, this is the case in most chronic diseases. Exercise and temperance will do infinitely more than medicine. By their means, most diseases may be overcome; but without them we may administer drugs as long as we please.

Voltaire sets this advice, which I have frequently inculcated, in so strong a light, that it may perhaps carry more conviction than any thing I can say. Ogul was a voluptuary, ambitious of nothing but good living: he thought that God had sent him into the world for no other purpose than to eat and drink: his physician, who had but little credit with him, when he had a good digestion, governed him with despotic sway, when he had eaten too much.

On feeling himself much and seriously indisposed by indolence and intemperance, he requested to know what he was to do, and the doctor ordered him to eat a basilisk, stewed in rose water, which he asserted would effect a complete cure. His slaves searched in vain for a basilisk; at last they met with Zadig, who was introduced to this mighty lord, and spoke to him in the following terms.

"May immortal health descend from Heaven to bless all thy days! I am a physician: at the report of thy indisposition, I flew to thy castle, and have now brought thee a basilisk, stewed in rose water. But, my lord, the basilisk is not to be eaten; all its virtue must enter through thy pores. I have enclosed it in a little ball, blown up and covered with a fine skin. Thou must strike this ball, with all thy might, and I must strike it back for a considerable time: and by observing this regimen for a few days, thou wilt see the effects of my art." The first day Ogul was out of breath, and thought he should have died with fatigue; the second he was less fatigued, and slept better. In eight days he recovered all the strength, all the health, all the agility and cheerfulness of his most agreeable years. Zadig then said unto him, "there is no such thing in nature as a basilisk; but thou hast taken exercise, and been temperate, and hast recovered thy health." In the same manner I say, that temperance and exercise are the two great preservers of health, and restorers of it when it is lost; and that the art of reconciling intemperance and health is as chimerical, as washing the Ethiopian white.

It will easily be perceived that the system of animal life which I have investigated, may be applied to all other general diseases, as well as the gout and those called nervous: I have merely given a view of these by way of specimen of its application.

Should these lectures contribute in any degree to lessen the future sufferings of my hearers, or any of their friends, I shall not have delivered them in vain. To be assured of this, would be the greatest pleasure that I could receive.

THE END.

From the Press of the Royal Institution of Great Britain,
Albemarle Street, London: W. Savage, Printer.

INDEX.

A. *Air*, its properties—its components *Animals*, specific temperature of *Appetite*, artificial *Arteries*, their structure and office *Assimilation*, from the blood *Attention*, fixed on new objects

B.
Banks, Sir Joseph, his almost fatal experience of cold
Beddoes, Dr. his remarks on temperature,
Bile, its properties
Blood, difference between arterial and venous
  contains iron
  changes produced on, by the different gases
  assimilation from
Bones, mechanism of
  structure of
Brown, Dr. John, his example followed
  declines a definition of excitability
  denies the existence of sedatives
  his cure of diseases of exhaustion objected to
  his theory will be as durable as Newton's philosophy
  not aware of the extent of his own theory

C. *Camera obscura Colour* of different nations *Cooper, Mr. Astley*, successfully perforates the tympanum of the ear *Circulation* of the blood through the lungs through the liver affected by centrifugal force discovered by Harvey, and derided *Cullen, Dr.* his system defective *Currie, Dr.* his treatment of typhus

D.
Darwin, Dr. quoted
Digestion, organs of
  process of
  diseases affecting
Diseases
  sthenic
  asthenic
  fallacy of symptoms
Diseases, method of cure
  spasmodic, of extreme vess.
  classification of
  nervous and bilious (so called)
  of the poor
Dollond, Mr. his achromatic lenses, first suggested by Dr. David Gregory

E.
Ear, description of
  diseases of
  tympanum perforated
Electricity, phenomena of
  useful in gout
Excitability, or vital principle
  how affected by heat, food, air, &c.
  accumulated by sleep
  three states of
  its accumulation, and exhaustion illustrated
  an unknown indefinable somewhat
  connected with oxygen
  hypothesis respecting this connexion
  scale of
  how to be regulated in gout
Exercise
  on horseback, best
Eye, description of

vertical section of
its accommodating power

F. *Food*, animal and vegetable in gout

G.
Galvanism, its identity with electricity
   accompanies oxydation
Gaming, &c. deplorable effects of, on the mind
Gases, their proportions in the atmosphere
   changes they produce on the blood
Gastric Juice, dissolves food
   proved by experiment
Gout
   denied to be salutary, or incurable
   not to be cured by drugs
   not hereditary
   mode of attack
   depends not on morbific matter
   its inflammation asthenic
   its symptoms and description
   its remissions accounted for
   the terms retrocedent, atonic, &c. misapplied
   mode of cure
   electricity useful
   Dr. Darwin's advice
   Diet and medicine in
Gravity, the circulation affected by it

H. *Habit*, its power *Harvey*, discovers the circulation of the blood, and is opposed *Health* previous to disease, investigated point of, on the scale *Heart*, description of *Heat*, its combinations animal, accounted for its effects on the system how it affects vegetables its debilitating effects exemplified *Hunger* and *thirst* Hunter, Mr. John, dissects torpedo, and electrical eel

I.
Inflammation, illustrated
   sthenic

asthenic
of gout asthenic
Iron, contained in the blood

L. *Lacteals*, described *Life*, principle of *Light*, its properties its exciting power *Lungs*, circulation through *Lymphatics* described

M. *Muscles*, mechanical action of *Music*, its effects

N.
Nerves, their distribution
Nervous and *bilious*, terms sometimes used without ideas
   diseases (so called)
   diseases really so
Newton, his query about elastic fluid misapplied
   even his conjectures important
   discovers the laws of sound
   his reason why the crystalline is densest in the middle
   knew not the cause of gravitation
Nonnaturals, misapplication of that term
Nosology

O. *Odours*, extreme subtilty of Haller, &c. fail in classing them convey nourishment *Opium*, its intoxicating effects its use in gout *Organs*, digestive *Oxygen*, experiments with connected with excitability

P. *Pain* encreases mental energy *Passions of the mind*, their effects *Pendulums*, laws of *Peristaltic motion Physiology*, its importance *Pulse*, the phenomena of

R.
Respiration, organs of
   how performed
   analogous to combustion
   chemically explained

S.
Schools, their temperature ill regulated
Sensation

  organs of
  more acute by pain
Senses, general account of
  external and internal
Skin
Sleep, accumulates excitability
Smell
  different in different animals
  Blumenbach's opinion on
  diseases of
Sound, production of
  media of
  illustrated
  velocity of
  musical
  reflection of
Spallanzani, his experiments on digestion
Spirituous liquors, their effects
  a singular mode of correcting the abuse of
Squinting, &c.
  method of cure
Stomach
  diseases of
Study, debilitating effects of

T. *Taste* diseases of *Temperance Temperature* of animals, specific and uniform *Touch* the basis of the other senses *Typhus*, Dr. Garnett's treatment of Dr. Currie's ditto

V.
Veins, their structure and office
Vision, manner of
  opinions on
  seat of
  why objects appear erect
  why seen single
  diseases and cure of
Voltaire, his story of Ogul, the epicure

W. *Water*, the best diluent *Wine*, its use and abuse

LIST OF SUBSCRIBERS.
L. s.
A.
Henry Ainslie, M.D. 10 10
James Peter Auriol, Esq. P.R.I. 5 5
James Acheson, Esq. *Oaks, Londonderry* 1 1
W. Dacre Adams, Esq. 1 1
Miss Adams 1 1
Mr. Adams, *Optician* 5 5
– – – Addington, Esq. 1 1
Mr. Thomas Addison, *Preston* 2 2
Mrs. Addison 2 2
John Aikin, M.D. 1 1
Mr. Arthur Aikin 1 1
Charles R. Aikin, Esq. *Surgeon* 1 1
William Alers, Esq. 2 2
Dr. Alexander, *Hallifax* 2 2
Henry Alexander, Esq. 1 1
E. Alexander, M.D. *Leicester* 1 1
William Allen, Esq. F.L.S. 2 2
Robert Anderson, Esq. 1 1
Dr. Anderson 1 1
Dr. Anderson, *Hammersmith* 1 1
A. Apsley, Esq. 1 1
Mr. D. Archer, *Bath* 1 1
Mr. J. T. Armiger 1 1
Nicholas Ashton, Esq. *Liverpool* 1 1
Mr. Thomas Aston 1 1
Abram Atkins, Esq. 3 3
William James Atkinson, Esq. 3 3
Mr. Accum 10 10
Mr. Noah Ayton 1 1

B.
Rt. Hon. Sir Joseph Banks, Bart. K.B.
  Pr. R.S. F.A.S. P.R.I. F.L.S. 10 10

Sir C. W. Rouse Boughton, Bart. P.R.I. 1 1
Sir William Blizard, F.R.S. F.A.S. 2 2
Mr. B. *Lancaster* 1 1
Dr. Babington, *Physician to Guy's Hospital.* 3 3
Miss Bacon 1 1
Miss Mary Bacon 1 1
John Bailey, Esq. 1 1
Andrew Bain, M.D. 5 5
John Baker, Esq. 10 10
William Baker, Esq. 10 10
Mr. Joseph Ball 1 1
Rev. Edward Balme 1 1
Mr. Bamford 1 1
Rev. John Banks, P.R.I. *Boston, Lincoln.* 1 1
Mr. Banks, *Leeds* 1 1
Charles Baratty, Esq. F.A.S. 1 1
Robert Barclay Esq. *Southwark* P.R.I. F.L.S. 1 1
Mr. William Barclay 1 1
R. J. Barlow, Esq. 3 3
Dr. Barrow 1 1
Rev. William Barton, *Whalley* 1 1
Mr. Barton 1 1
Thomas Baskerfeild, Esq. F.A.S. 1 1
Jacob Bath, Esq. *Surgeon, 66th Reg.* 1 1
Robert Batty, M.D. P.R.I. F.L.S. 5 5
Mr. Charles Baumer 1 1
E. Bayley, Esq. *St. Petersburg* 1 11 6
Mr. John Baynes 1 1
Mr. Beckett 1 1
Hugh Bell, Esq. 1 1
William Bell, Esq. *St. Petersburg* 1 11 6
Mr. Bell 5 5
Christopher K. Bellew, Esq. 2 2
Rev. N. Benezer, *Swaffham* 1 1
Nathaniel Barnardiston, Esq. 1 1
Thomas Bernard, Esq. P.R.I. 5 5

W. Berwick, Esq. *St. Petersburg* 1 11 6
Richard Best, Jun. Esq. 1 1
Henry Bickersteth, Esq. *Kirkby Lonsdale* 2 2
Henry Bickersteth, Jun. Esq. *Kirkby Lonsdale* 1 1
Mr. John Bickersteth 2 2
Mr. E. Bickersteth 1 1
T. Bigge, Esq. *Newcastle upon Tyne* 1 1
Mr. George Biggin 1 1
W. Bindsale, Esq. *Pickering, Surgeon*
Charles Binney, Esq. 1 1
Joseph Birch, Esq. M.P. 3 3
Dr. Birkbeck 5 5
Morris Birkbeck, Jun. Esq. *Wanborough, Surry* 1 1
Thomas Blacker, Esq. P.R.I. 5 5
Mr. Blades 1 1
Alexander Blair, Esq. P.R.I. 5 5
William Blair, Esq. 3 3
Dr. Blake, *Taunton* 1 1
Dr. Blane 5 5
Mr. Henry Blatch, *Surgeon* 1 1
Ditto, *2nd Subscription* 4 4
Ralph Blegborough, M.D. P.R.I. 2 2
John Bliss, Esq. 1 1
Robert Blundell, Esq. 1 1
Samuel Boddington, Esq. P.R.I. 1 1
George Bodley, Esq. 1 1
Thomas Bodley, Esq. 1 1
John Bolton, Esq. 1 1
Dr. Bolton 5 5
Thomson Bonar, Esq. P.R.I. 10 10
George Booth, Esq. 2 2
John Bostock, M.D. 2 2
Josiah Boydell, Esq. 3 3
Daniel Braithwaite, Esq. F.R.S. F.A.S. 5 5
Miss Braithwaite 2 2

Mr. James Braithwaite 1 1
— — —Brande, Esq. 1 1
Joseph Brandreth, M.D. 5 5
Mr. Brandreth, *Surgeon* 3 3
Mr. Brandreth, *Attorney* 2 2
Messrs. Brash and Reid, *Glasgow* 2 2
John Breare, Esq. 1 1
Martin Bree, Esq. 1 1
J. H. Brehmer, Esq. *St. Petersburg* 1 11 6
Brentford Book Society 1 1
William Bridgman, Esq. P.R.I.
  F.L.S. 1 1
Mr. Richard Briggs 2 2
Mrs. Richard Briggs 1 1
Lowbridge Bright, Esq. *Bristol*
British Library, *St. Petersburg* 1 1
Theodore Henry Broadhead, Esq.
  F.A.S. P.R.I. 2 2
Messrs. Broderip 5 5
James Brogden, Esq. M.P. 3 3
Mr. E. Brook 1 1
Joshua Brooks, Esq. *Lecturer on
  Anatomy, Surgery, &c.* 5 5
James Brown, Esq. *Leeds* 1 1
J. Brown, M.D. *Islington* 1 1
Mr. Brown 1 1
— — —Bryant, Esq. *Lynn Regis,
  Surgeon* 1 1
Dr. Buchan 1 1
Rev. Gilbert Buchanan, LL.D.
Rector of Woodmanstow, Surry 1 1
Mr. Buchenson, *Surgeon* 1 1
Buckingham Book Society 1 1
George Buckle, Esq. 2 2
William Buckley, Esq. 1 1
Mr. Henry Budd 1 1
Mr. John Buddle 1 1
Mr. J. Buncomb, *Taunton, Somerset.* 1 1
Mr. A. S. Burkitt, *Chemist* 3 5 6

Charles Burney, LL.D. F.R.S.
  F.A.S. P.R.I. *Greenwich* 1 1
Robert Burns, Esq. 1 1
Mrs. Butts 1 1

C.
Sir John Chetwode, Bart. 1 1
Henry Cavendish, Esq. F.R.S.
  F.A.S. P.R.I. 10 10
Thomas Coutts, Esq. P.R.I. 10 10
Mr. John Callow 1 1
Dr. Alexander Campbell, *Secretary
  to the Medical Board, Calcutta* 1 1
Dr. Campbell, *Lancaster* 1 1
Colonel James Campbell 1 1
Mr. Card 1 1
Anthony Carlisle, Esq. F.L.S. 1 1
Mr. John Carpenter, Senr. *Lyme*
J. C. Carpue, Esq. 2 2
George Carr, Esq. *St. Petersburg* 1 11 6
Thomas W. Carr, Esq. P.R.I. 1 1
Mr. W. Cary 1 1
William Cass, Esq. 1 1
Dr. Cassels, *Lancaster* 1 1
D. Cassidy, Esq. 1 1
John Cates, Esq. 1 1
H. Cayley, Esq. *St. Petersburg* 1 11 6
William Chamberlain, Esq. 1 1
L. B. Chateauneauf, Esq. 1 1
Chester Public Library 1 1
Henry Chevalier, Esq. F.L.S. 2 2
Mrs. Chevalier 1 1
James Chew, M.D. *Blackburn* 1 1
Mr. Christian, *Attorney* 1 1
— — —Clapham, Esq. *Thorney
  Abbey, Surgeon* 1 1
John Clarke, Esq. 1 1
J. Calvert Clarke, Esq. 5 5
Ralph Clarke, Esq. 1 1

Richard Hall Clarke, Esq. *Devon.* 1 1
Mr. John Clay 1 1
S. P. Cockerell, Esq. P.R.I. 2 2
Miss Codrington 2 2
Mr. John Cohen 1 1
Mr. Collier 1 1
Mr. Coltman, *Surgeon* 2 2
Mr. Colton 1 1
Mr. Cooke 1 1
B. Coombe, Esq. 2 2
Astley Cooper, Esq. F.R.S. 5 5
Rev. Stuart Corbett
Adam Cottam, Esq. *Whalley, near Blackburn* 1 1
Mr. Cotter, *Godstone, Surry* 1 1
Mr. Cotter 1 1
John Cowan, Esq. 1 1
Theodore Cox, Esq. 2 2
John Craig, Esq. *Glasgow* 1 1
Mr. Cregg 1 1
Thomas Creser, Esq. 1 1
William Cresswell, Esq. 1 1
Alexander Crichton, M.D. F.R.S. F.L.S. 2 2
Mr. Crigg 1 1
George Crooke, Esq.
Hugh Cross, Esq. 2 2
C. Cuppage, Esq. 1 1
Mr. William Cuppage 2 2
James Currie, M.D. F.R.S. *Physician to Guy's Hospital* 2 2
— — —Currie, M.D. 5 5
— — —Cuthbert 1 11 6
Mr. John Cuthbertson 1 11 6

D.
Sir George Duckett, Bart. P.R.I. 1 1
Peter Denys, Esq. P.R.I. 10 10
Lady C. Denys 5 5

George William Denys, Esq. P.R.I. 5 5
David Dale, Esq. 5 5
J. P. Dale, Esq. 5 5
Thomas Dale, M.D. 1 1
Mr. Daniel Dale 5 5
Mr. Daltera, *Attorney* 1 1
Mr. Darbyshire 1 1
Mr. Davie, *St Thomas's Hospital* 1 1
Dr. Davies, *Bath* 1 1
Dr. Davison, *Harewood*
Humphry Davy, Esq. F.R.S. 5 5
Mr. Joseph Dawson 1 1
Thomas Dawson, Esq. *Surgeon* 1 1
Mr. Thomas Dawson 1 1
Mr. Dawson 2 2
Thomas Day, Esq. P.R.I. 5 5
Miss Day 2 2
Mrs. Delamore 1 1
Thomas Demson, Esq. 1 1
R. Dennison, M.D. and F.A.S. 2 2
Osbert Denton, Esq. *Chelsea* 10 0
Mrs. Denton 10 0
Miss Denton 5 0
Mr. Benjamin Devaynes 1 1
N. Devy, Esq. 1 1
Charles Dibdin, Esq. 5 5
William Dick, Esq. 1 1
W. Dickson, LL.D. *Clipstone St.* 2 2
Charles Dilly, Esq. 2 2
Mr. William Dinsley, *Leeds* 1 1
— — —Doig, Esq. 1 1
Mr. Doig 1 1
George Dominicus, Esq. 2 2
Rev. Thomas Donald, *Anthorne, Cumberland* 1 1
Andrew Douglas, Esq. F.R.S. P.R.I. 5 5
Mr. James Dove 1 1
Henry Downer, Esq. 1 1
Mr. Drewe 1 1

F. Duboulay, Esq. 1 1
Dr. Andrew Duncan, *Edinburgh*
Dr. Andrew Duncan, Jun. *Do.*
Mrs. Dunlop, *Hammersmith* 1 1
Miss Dunlop, *Hammersmith* 1 1
James Dupre, Esq. P.R.I. 1 1

E.
Marquis of Exeter, F.R.S. F.A.S. 5 5
Marchioness of Exeter 5 5
The Earl of Egremont, F.R.S.
  F.A.S. P.R.I. 10 10
Sir H. C. Englefield, Bart. F.R.S.
  F.A.S. P.R.I. 3 3
Sir James Earle 2 2
Hon. William Elphinstone 1 1
R. G. Ealand, Esq. 1 1
Willis Earle, Esq. 2 2
— — —Eccles, Esq. *Clithero* 1 1
D. B. Plantagenet Eccleston, Esq. 1 1
Library of the University of
  Edinburgh
Mr. Edmondson, *Surgeon, Lancaster* 1 1
Mr. Richard Elgar 1 1
George Ellis, Esq. F.R.S. F.A.S.
  P.R.I. 1 1
John Fullerton Elphinstone, Esq.
East Lodge, Enfield 1 1
Alexander Erskine, Esq. F.L.S. 1 1
George Erving, Esq. 1 1
John Esdaile. Esq. P.R.I. 1 1
Isaac Espinasse, Esq. 2 2
William Etty, Esq. 1 1
Mr. Evans 1 1

F
Col. the Hon. T. W. Fermor, P.R.I. 10 10
James Fallofeild, Esq. P.R.I. 2 2
Vere Fane, Esq. 1 1

Thomas Farquhar, Esq. 1 1
Robert Farquhar, Esq. *Rochester* 1 1
Mr. Faulder 1 1
W. Fawkes, Esq. *Farnley, near Leeds* 3 3
John Feltham, Esq. 2 2
Richard Firmin, Esq. 3 3
Mr. M. Fitzgerald 1 1
Lewis Flannagan, Esq. 2 2
Arch. Fletcher, *Advocate, Glasgow* 2 2
Caleb Fletcher, Esq. 2 2
Mr. Benjamin Flockton
— — —Forbes, Esq. 1 1
Henry Fock, Esq. *St. Petersburg* 1 11 6
Mr. Ford, *Pimlico* 1 1
— — —Ford, Esq. 2 2
R. Forrester, Esq. *St. Petersburg* 1 11 6
B. M. Forster, Esq. 1 1
Rev. Edward Forster, F.R.S. F.A.S. P.R.I. 1 1
Thomas Foster, Esq. 1 1
Thomas Fothergill, Esq. 2 2
John Foulks, Esq. 2 2
Joseph Fox, Esq. 2 2
Rev. M. Foxcroft 1 1
James Fraser, Esq. P.R.I. 1 1
F. Freelong, Esq. 1 1
— — —French, Esq. 1 1
A Friend 2 2
A friend, W. M. 1 1
A friend, J. C. 1 1
A friend, T. K. 2 2
A friend, E. 4 4
A friend, T. 10 0
A friend, J. G. 5 5
A friend, A. 2 2
A friend, R. W. by Mr. Lawson 1 1
A friend, W. W. 1 1
A friend, I. A. W. 1 1
A friend, I. S. 1 1

A friend, A. B. 2 2
A friend, I. B. 5 5
A friend, W. D. 2 2
A friend, D. D. 2 2
A friend, J. L. Esq. 3 3
A friend, E. A. 1 1
A friend, I. S. 1 1
A friend, I. B. 1 1
A friend, M. W. 1 1
A friend, M. B. 5 0
A friend, I. A. W. 1 1
A friend, E. D. 1 1
A friend, J. D. 1 1
A friend, J. P. D. 1 1
A friend, S. B. D. 4 4
A friend, S. D. 5 5
A friend, Mr. B. 1 1
A friend, Mr. B. 1 1
A friend, S. E. D. 1 1
A friend, S. E. D. 1 1
A friend, S. M. D. 1 1
A friend, I. T. by Mr. I. J. Armiger 1 1
A friend, by Dr. Batty 3 3
A friend, I. H. D. by Dr. Batty 3 3
A friend, by D. Braithwaite, Esq. 3 3
A friend, by Miss Braithwaite 2 2
A friend, by Mr. Gear 1 1
A friend, H. L. by Mr. Thomas 1 1
A friend, S. H. by Mr. Lawson 2 2
A friend, by Mr. Rathbone 5 5
Edmund Fry, M.D. 1 1

G
Sir James Whalley Gardiner, Bart.
Clerk Hill, near Blackburn 3 3
Mr. William Gaitskell 3 3
Peter Garforth, Esq. 5 5
John Gaythorne, Esq. 2 2
Robert Gear, Esq. 2 2

Do. for a friend, *no Book* 1 1
John Geddes, Esq. 2 2
James Gerard, Esq. 3 3
Mr. Robert Gibson 2 2
Mr. W. Gilbert, *Optician* 2 2
Collin Gillespie, Esq. 1 1
William Gillespie, Esq. 5 5
Robert Gillespie, Esq. 1 1
Thomas Gisborne, M.D. F.R.S. 1 1
Henry Glossop, Esq. *Cambridge* 2 2
Mrs. Gooch 1 1
— — —Good, Esq. 1 1
Mr. Good 1 1
Mr. Alexander Gordon, *Surgeon* 1 1
Mr. Alexander Caudleraig Gordon, *Edinburgh* 1 1
Mrs. Maxwell Gordon, *Edinburgh* 2 2
Thomas T. Gorsuch, Esq. 1 1
Mr. Samuel Gosnell, *Printer* 1 1
Joseph Gough, Esq. 1 1
Archibald Graham, Esq. *Glasgow* 1 1
Robert Graham, Esq. 1 1
Mrs. Henry Grant 2 2
Alex. Grant, Esq. *St. Petersburg* 1 11 6
Edward Whitaker Gray, M.D. Sec. R.S. 1 1
R. Greaves, M.D. *Reading, Berks.* 2 2
Mr. Griffenhof, *Hampton* 1 1
Thomas Green, Esq. 3 3
Mrs. Green 2 2
— — —Greenwood, Esq. *Lancaster, Surgeon* 1 1
John Grenfell, Esq. *Chelsea* 1 1
Mr. Griffin 1 1
T. Grimston, Esq. *Kilnwick, York.* 1 1
Henry Grimston, Esq. F.L.S. 1 1
John Guthrie, Esq. 2 2

H

Sir John Cox Hippisley, Bart.
F.R.S. F.A.S. P.R.I. 5 5
Hon. Mr. Justice Heath, P.R.I. 1 1
Messrs. Hoare 21 0
Mr. Hadwen, *Lancaster* 1 1
G. H. C. Hahn, Esq. *Wimbledon* 1 1
Samuel Hailstone, Esq. 1 1
John Hale, Esq. P.R.I. 2 2
Mr. T. Hall 2 2
Mr. Halliday, *Everton, near Liverpool* 1 1
Dr. James Hamilton, *Professor of Midwifery, University, Edinburgh*
Mr. John Hamilton 1 1
Colonel Hamilton, *His Majesty's Consul, Norfolk, Virginia* 5 0
G. H. Hancock, Esq. 1 1
Mr. Hankey 5 5
Mr. John Harding 1 1
Robert Hardy, Esq. *Whalley, near Blackburn, Surgeon* 1 1
Mr. John Hardy 1 1
Rev. Richard Harrington, *Hayley, Worcestershire* 1 1
Rev. Robert Harris, *Preston* 1 1
J. Harrison, Esq. *Kirkby Lonsdale* 5 5
Thomas Harrison, Esq. 1 1
Mr. Joseph Harrison, *Bury* 3 3
Mr. William Harrison, *Manchester* 1 1
Harrowgate Book Club 1 1
Enoch Harvey, Esq. *Liverpool* 2 2
Joseph Haskins, Esq. P.R.I. 3 3
William Huskisson, Esq. 1 1
Charles Hatchett, Esq. F.R.S.
F.L.S. P.R.I. 10 10
Dr. Hawes, (W. H. S.) 1 1
Mr. B. Hawes 1 1
Richard Haworth, Esq. 2 2

Mr. Joseph Hawthorne, *Reading* 1 1
Thomas Hay, Esq. 1 1
— — — Haygarth, M.D. *Bath* 2 2
William Headington, Esq. 1 1
John Heath, M.D. *Fakenham, Nor.* 1 1
Mr. Richard Heathfield 1 1
Mr. Henbest 2 2
William Henderson, Esq. 2 2
Messrs. Thomas and William Henry 5 5
Messrs. T. and W. Henry 5 5
Benjamin Arthur Heywood, Esq. 2 2
B. A. Heywood, Esq. 2 2
Mr. Hicks, *Surgeon* 1 1
Dr. Higgins 2 2
John Benton Higgon, Esq. 1 1
Henry Hinckley, Esq. 2 2
John Hinckley, Esq. P.R.I. 2 2
Elias Hocker, Esq. *Brook House, Devon.* 1 1
Hoddesdon Book Society 2 2
Mrs. Hodge 1 1
Henry Hodgson, Esq. 1 1
Rev. Dr. Hodgson, *Market Rasen, Lincolnshire* 1 1
Charles Holford, Esq. *Hampstead* 1 1
Mr. W. Holland 1 1
Edward Holme, M.D. 2 2
Mr. Jacob Holme 1 1
Everard Home, Esq. F.R.S. P.R.I. 5 5
Mr. Hommey, *Military Academy, Woolwich* 1 1
Mr. Hooke, *Optician*, P.R.I. 2 2
Dr. Hooper, F.L.S. 5 5
Dr. Thomas C. Hope, *Edinburgh*
Rev. J. J. Hornby 1 1
Mr. J. D. Hose, Jun. 1 1
Mr. Houseman 1 1
Luke Howard, Esq. 3 3
S. R. Howard, Esq. 1 1

T. M. Hubert, Esq. 1 1
Henry Hughs, Esq. 3 3
Richard Hughes, Esq. 5 5
Hull Subscription Library 1 1
John Hull, M.D. 2 2
Dr. Hull 2 2
Dr. Hulme 2 2
N. Hulme, M.D. F.R.S. and F.A.S 3 3
Mr. N. Hume 3 3
Mr. J. Hume 1 1
Mrs. Hunt 3 3
Robert Hunter, Esq. *Kew* 1 1
Dr. Hunter, *York*
Mr. John Hustler 1 1
Walter Hutton, Esq. 2 2
Mr. Hutton, *Sheffield*

J
G. Jackson, Esq. *St. Petersburg* 1 11 6
Mr. Jackson, *Lancaster* 1 1
Miss Sarah Jackson 1 1
William James, Esq. 1 1
Thomas I'anson, Esq. 2 2
— — —Jaques, M.D. *Harrowgate* 2 2
— — —Jardine, M.D. 2 2
George Jeffery, Esq. 1 1
Edward Jenner, M.D. F.R.S. F.L.S. 10 10
Edward Jenings, Esq. *Kensington* 1 1
William Ingham, Esq. *Newcastle upon Tyne* 2 2
Rt. Innes, Esq. *Newcastle upon Tyne* 1 1
Edward Johnson, Esq. *Mile End* 1 1
Christopher Johnson, Esq. 1 1
R. Johnson, Esq. 1 1
— — —Jones, Esq. *Chelsea* 1 1
W. Jones, Esq. *St. Petersburg* 1 11 6
Thomas Jones, Esq. *Llandisillio Hall, Oswestry* 1 1
Mr. John Jones, *Surgeon* 2 2

Stephen Jones, Esq. 1 1
Gibbs Walker Jordan, Esq. F.R.S. 3 3
Major Jourdan 1 1
Rev. J. Joyce 1 1
Thomas Irving, Esq. 2 2
Thomas Irwin, Esq. 1 1
Mr. Ives, *Surgeon, Chertsey*

K
The Hon. George Knox, F.R.S. 10 10
Frederick Kanmacher, Esq. P.R.I.
  F.L.S. 1 1
Mr. B. A. Keck, *Leeds*
Miss Keene 2 2
Mr. P. Kelly 1 1
Dr. E. Kentish, *Bristol*
Mr. Jonathan Key 1 1
T. King, Esq. *Carshalton, Surry* 1 1
James King, Esq. 2 2
J. King, Esq. *Finsthwaite, Kendal* 1 1
Messrs. Knipe and Mower 2 2
Kirkby Lonsdale Book Club 1 1
John Knox, Esq. 1 1
William Knox, Esq. 1 1

L
Right Hon. Lady Lyttelton 3 3
Sir James Lake, Bart. 1 1
George Lackington, Esq. 2 2
Mr. W. Laforest 2 2
H. Lamatte, Esq. 1 1
Mrs. Lappan 1 1
Samuel Lawford, Esq. P.R.I. *Peckham* 1 1
Thomas Wright Lawford, Esq. 1 1
Mr. Samuel Lawford, jun. 1 1
Johnson Lawson, Esq. P.R.I. 5 5
Henry Lawson, Esq. 5 5
Rev. G. Lawson 1 1
James Inglish Lawson, Esq. 1 1

William Leake, Esq. 2 2
Mr. Joseph Leay 1 1
Mr. William Leay 1 1
Rev. Francis Lee 2 2
John Leeds, Esq. *Chelsea* 1 1
James Leighton, Esq. 2 2
Dr. Lempriere 1 1
J. Lestourgeon, Esq. 1 1
Thomas Lewis, Esq. P.R.I. 2 2
William Lewis, Esq. F.L.S. 5 5
Mr. Lewthwaite 1 1
Library of the Sheffield Infirmary
Henry Lider, Esq. *Stilton* 1 1
John Lightbody, Esq. *Liverpool* 5 5
A. S. Lillingston, Esq. *Llyne* 1 1
Lincoln Medical Society
Literary Fund 30 0
Literary Society at Newcastle 2 2
Mr. Littlewood 1 1
Liverpool Library 1 1
John Lloyd, Esq. *Bashall Lodge, Yorkshire* 1 1
H. E. Lloyd, Esq. 1 1
Mr. James Long, *Optician* 3 3
C. Lormar, Esq. *Bath* 1 1
Mr. R. Low, *Surgeon* 1 1
James Lowe, Esq. 2 2
Mrs. W. Lowry 5 5
Captain John Lucas 1 1
Mr. Lucas 2 2
James Ludlam, Esq. *Homerton* 1 1
Mr. Lumley 1 1
William Luxmoore, Esq. P.R.I. 1 1
T. Luxmoore, Esq. 1 1
Lynn Subscription Library 1 1
John Lyon, M.D. 3 3

M
Right Hon. Lord Muncaster

John Mac Arthur, Esq. P.R.I. 1 1
John McCartney, M.D. 2 2
A. McCausland, Esq. *St. Petersburg* 1 11 6
Angus McDonald, Esq. 2 2
Angus McDonald, M.D. 1 1
Dr. Macdonald *Taunton, Somerset.* 1 1
Duncan Macfarlane, Esq. 5 5
George McIntosh, Esq. 4 4
Dr. McIntosh 2 2
Mr. Mackender 1 1
Alexander McLeay, Esq. 1 1
Mr. Maddon, jun. 1 1
John Mair, Esq. 2 2
Mr. Malcolm 1 1
F. H. Maltz, jun. Esq. *Hamburgh* 1 1
Alexander Marcet, M.D. P.R.I. *Assistant Physician to Guy's Hospital* 2 2
Charles Marsh, Esq. F.A.S. P.R.I. 1 1
Dr. Marshal 1 1
Mr. William Marshall 1 1
William Maud, Esq. 1 1
P. W. Mayow, Esq. 1 1
W. W. Mayow, Esq. 1 1
Samuel Mellish, Esq. P.R.I. 1 1
W. Menish, Esq. P.R.I. *Bromley, Kent* 2 2
Richard Meux, jun. Esq. 5 5
John Middleton, Esq. 1 1
John Milford, Esq. 2 2
James Millar, Esq. *College, Glasgow* 1 1
Captain Millar 1 1
Miss Millar 1 1
— — —Miller, Esq. *Peterborough, Surgeon* 1 1
James Milne, *College, Glasgow* 2 2
Charles Minier, Esq. 2 2
William Minier, Esq. 2 2
Mr. Minchell, *Surgeon* 3 3

Ed. Moberly, Esq. *St. Petersburg* 1 11 6
Mr. Mogg 1 1
Dr. Alex Monro, Sen. *Edinburgh*
Matthew Montagu, Esq. F.R.S. 1 1
Charles Montague, Esq. *Surgeon* 1 1
Robert Monteith, Esq. 1 1
Jon Monteith, Esq. 1 1
Robert Montgomerie, Esq. 1 1
G. Morgan, Esq. *Norfolk, Virginia* 5 0
Mr. Joseph Morgan 1 1
William Morley, jun. Esq. P.R.I. 1 1
John Morris, Esq. 2 2
Mr. James Moss 1 1
P. S. Munn, Esq. 1 1
Sir John Chardin Musgrave, Bart. 1 1

N
Edward Nairne, Esq. F.R.S. P.R.I. 5 5
Rev. Archdeacon Nares, F.A.S. 2 2
Mr. Charles Newby 2 2
Mr. J. Nicholas 1 1
George Nicholson, Esq. *Ponchnill, near Ludlow* 1 1
Mr. Francis Nicholson, *Artist, Somers Town* 2 2
Mr. William Nicholson 5 5
Mrs. Norman, *Chelsea* 1 1
William Norris, Esq. 2 2
Henry Norris, Esq. *Davy-Hulme-Hall, near Manchester* 5 5
Mr. Norris 5 5
William North, Esq. 1 1
Thomas Northmore, Esq. 5 5
Messrs. Norton and Son, *Bristol*

O
James Ogilvy, Esq. 1 1
Edward Ogle, Esq. 1 1
Henry Okey, Esq. 1 1

Mr. Oliver 1 1
Mr. Serjeant Onslow 1 1
Rev. Richard Ormerod, *Kensington* 1 1
Alexander Oswald, Esq. 5 5
R. Oswin, Esq. 2 2
Mr. Otley 1 1
— — —Owen, Esq. 10 10

P
Royal College of Physicians 20 0
The Earl of Pomfret, P.R.I. 10 10
Sir William Pepperell, Bart. 2 2
Sir Christopher Pegge, Knt. *Regius Professor of Physic, Oxford*
Philip Palmer, Esq. 5 5
Felix Palmer, Esq. 3 3
Mrs. Palmer 3 3
Executors of Mrs. Palmer 10 0
Thomas Park, Esq. F.A.S. 1 1
Mr. Park, *Surgeon* 3 3
Mr. J. C. Parker 5 5
Thomas Parker, Esq. 5 5
Mr. Robert Parker 1 1
G. Parkinson, Esq. 3 3
— — —Parkinson, Esq. *Clitheroe* 1 1
Dr. Parkinson, *Lancaster* 1 1
John Parry, Esq. 5 5
John Pattison, Esq. 7 7
Mrs. Paulin, *Chelsea* 1 1
William Payne, Esq. *Bradford* 1 1
Thomas Paytherus, Esq. 2 2
John Peake, Esq. *Battersea* 1 1
C. Pears, Esq. F.L.S. 2 2
Rev. William Pearson, P.R.I. 3 3
William Pearson 3 3
Thomas Pearson, Esq. 1 1
William Scott Peckham, Esq. 1 1
Edmund Peckover, Esq. *Bradford* 1 1
W. H. Pepys, Esq. 5 5

W. H. Pepys, jun. Esq. P.R.I. 5 5
Edmund Pepys, Esq. 1 1
Thomas Perceval, M.D. F.R.S. F.A.S. 2 2
Dr. Percival 2 2
Mr. B. D. Perkins 1 1
W. Perrin, Esq. 2 2
James Perry, Esq. P.R.I. 3 3
Louis Hayes Petit, Esq. F.A.S. 2 2
Mr. C. H. Pfeffel 1 1
Rev. W. G. Phillips, P.R.I. 1 1
William Phillips, Esq. 1 1
Mr. R. Phillips, *Bookseller* 5 5
Mr. Richard Phillips 1 1
Mrs. Charles Phillott, *Bath* 1 1
Johnson Phillott, Esq. *Bath* 1 1
Philosophical Society, *Hythe, Kent* 1 1
Physical Society, *Guy's Hospital* 6 6
Library of the College of Physicians, Edinburgh
J. K. Picard, Esq. *Summergang's House, near Hull* 2 2
Charles Pieschell, Esq. P.R.I. 2 2
Gillery Pigott, Esq. 2 2
T. Pilkington, Esq, *Bewdley, Surgeon* 1 1
William Pilkington, Esq. 2 2
Mrs. Pilkington 1 1
William Pim, Esq. 2 2
I. Pinchard, Esq. *Taunton* 1 1
Dr. Pinckard 2 2
Thomas Pitt, Esq. F.A.S. P.R.I. 2 2
William Plasted, Esq. *Chelsea* 1 1
— — —Pocock, Esq. 1 1
Plymouth Literary Society 1 1
R. Podmore, Esq. 1 1
Mr. William Pollard, *Bradford* 1 1
Thomas C. Porter, Esq. 1 1
Portsmouth Book Society 2 2
The Rev. Josiah Pratt, F.A.S. 1 1

J. Prescott, Esq. *St. Petersburg* 1 11 6
The Rev. I. Preston, *Flasby Hall, Yorkshire* 1 1
Joseph Price, Esq. 1 1
Rev. William Price, *Bath* 1 1
Benjamin Price, Esq. 1 1
Mr. Joseph Priestly, *Bradford* 1 1
Mr. Prince 1 1
Mr. Pritt 1 1
Mr. Pullen 1 1
T. Pulley, Esq. 1 1

R
The Royal Institution of Great Britain 50 0
Earl of Radnor, F.R.S. F.A.S. 4 0
Hon. Dr. Francis Rigby, *Jamaica*
The Hon. Mrs. Rollo 2 2
Rev. Thomas Rackett, F.R.S. F.A.S. P.R.I. F.L.S. 1 1
Mrs. Rackett 1 1
Miss Rackett 0 10 6
Richard Radford, Esq. 1 1
Rev. Dr. Raine, F.R.S. F.A.S. 2 2
Richard Raley, Esq. *Clare Hall, Cambridge* 1 1
George Ranking, Esq. P.R.I. 2 2
John Ranking, Esq. *St. Petersburg* 1 11 6
Mr. Rathbone 5 5
Thomas Rawson, jun. Esq. 2 2
Mr. W. L. Reay, *Surgeon* 2 2
Rev. Dr. Rees, F.R.S. 1 1
Joshua Reeve, Esq. P.R.I. 5 5
T. R. Reid, Esq. 1 1
James Remnant, Esq. 2 2
Thomas Renwick, M.D. 3 3
George Reveley, Esq. 1 1
S. W. Reynolds, Esq. 2 2
Dr. Reynolds, F.R.S. F.A.S. 2 2

D. Ricardo, Esq. 1 1
Mr. Rich, *Fryston, Yorkshire* 5 0
Mrs. Richards 5 5
Isaac Ried, Esq. 1 1
J. Ring, Esq. 2 2
John Risdon, Esq. *Peckham* 1 1
Mr. Rittson, *Lancaster, Surgeon* 1 1
Thomas Roberts, Esq. 1 1
James Robertson, Esq. 1 1
Arthur Robinson, Esq. 1 1
Mr. Robinson, *York* 1 1
— — —Robinson, Esq. 1 1
Jonathan Rogers, M.D. 3 3
Dr. Rogerson 2 2
John Rollo, M.D. 1 1
Thomas Rolph, Esq. *Peckham* 1 1
Samuel Romilly, Esq. 2 2
William Rosco, Esq. 3 3
Dr. Rowley 5 5
— — —Rowley, M.D. 5 5
J. Rowlat, Esq. *St. Petersburg* 1 11 6
Edward Rudge, Esq. F.L.S. 1 1
Rev. Joshua Ruddock, A.M. 2 2
Mr. L. T. Rupp 2 2
Mr. Rupp 2 2
T. Rutt, Esq. 1 1
John Rutter, M.D. 2 2

S
Earl Stanhope, F.R.S. P.R.I. 1 1
Sir Richard Joseph Sullivan, Bart.
 F.R.S. F.A.S. P.R.I. 5 5
Sir John St. Aubyn, Bart.
Mr. William Salter 1 1
Dr. Satterley, *Tunbridge Wells* 1 1
George Saunders, Esq. F.A.S. 1 1
Thomas Scott, Esq. 1 1
Mrs. P. Scott 2 0
Mr. G. M. Sears 1 1

John Semple, Esq. 2 2
Miss Senior, *Sittingbourne* 1 1
Thomas Serjeant, Esq. 1 1
Dr. Serney, *Boston, Lincolnshire* 1 1
Henry Settrey, Esq. 1 1
Mr. Edmund Shadwick 1 1
George Grinley Sharpe, Esq. *Peckham, Surgeon* 1 1
Mr. Shaw, *Surgeon* 2 2
Townley Rigby Shawe, Esq. *Preston* 1 1
Miss Shee 1 1
Miss E. Shee 1 1
Miss J. Shee 1 1
Miss H. Shee 1 1
Library of Sheffield Infirmary
W. Shepherd, Esq. 1 1
Stephen Shute, Esq. 1 1
Mr. William Silver, *Portsmouth* 1 1
Richard Simmons, Esq. *Surgeon* 2 2
Mr. John Simmons 1 1
Mr. Thomas Simpkin 1 1
Mr. John Simpson 1 1
Mr. Robert Simpson, jun. 1 1
Samuel Everingham Sketchley, Esq. 1 1
Mr. Alexander Sketchley 1 1
Mr. Alexander Sketchley, *Clapham* 1 1
— — —Slade, Esq. 1 1
R. Smales, Esq. 1 1
Rev. Mr. Smirnove 1 1
John Stafford Smith, Esq. 1 1
J. R. Smith, Esq. 5 5
Nathan Smith, Esq. 1 1
— — —Smith, Esq. *Lancaster, Surgeon* 1 1
James Smith, Jun. Esq. *Glasgow* 2 2
Dr. Smith 1 1
Rev. Thomas Jenys Smith *Dulwich* 1 1
Mr. Egerton Smith 1 1
Mr. Frederick Smith 1 1

Mr. Smithson, *Heath, Wakefield* 2 2
Society for the Encouragement of
  Arts, Manufactures, and Commerce 5 5
Samuel Solly, Esq. P.R.I. 2 2
Archibald Sorley, Esq. 2 2
James Sowerby, Esq. *Lambeth* 1 1
John Spence, Esq. *Greenock* 3 3
Knight Spencer, Esq. 2 2
— — —Spencer, Esq. *Surgeon* 2 2
John Spencer, Esq. 2 2
John Sperling, Esq. 1 1
M. Spratt, Esq. 1 1
William Stanistreet, Esq. 3 3
William Stanniforth, Esq. *Sheffield, Surgeon*
Rev. T. Starkie, A.M. *Blackburn* 1 1
S. F. Statham, Esq. *Arnold, Notting.* 1 1
Mr. William Standley, *Bradford* 1 1
Mr. Thomas Steel 1 1
J. Steers, Esq. 1 1
Francis Stephens, Esq. 1 1
Mr. J. Stephenson 1 1
Rev. W. Stevens 1 1
William Stodart, Esq. P.R.I. 5 5
Mr. James Stodart, P.R.I. 5 5
Mr. M. Stodart 1 1
Miss Strickland, *Boynton*
Sunderland Subscription Library 1 1
John Swale, Esq. 1 1
Mr. Swan 1 1
Mr. Sydenham, *no Book* 1 1
Godfrey Sykes, Esq. 1 1

T
Peter Tahourdin, Esq. 1 1
John Tate, jun. Esq. 1 1
C. H. Tatham, Esq. 1 1
B. Tayle, Esq. 1 1
John Taylor, Esq. 3 3

James Taylor, Esq. 2 2
R. S. Taylor, Esq. 1 1
Mr. I. Taylor, *Gomersall*
Arthur Tegart, Esq. 1 1
Dr. Temple 5 5
Mrs. Tennant 2 2
Mr. Frederick Thackerey, _Emanuel College, Cambridge 1 1
H. L. Thomas, Esq. 1 1
Robert Thompson, Esq. F.A.S. 2 2
Robert Thompson, jun. Esq. 1 1
Mrs. Thompson 2 2
John P. Thompson, Esq. 1 1
Messrs. Thomson and Son, *Manchester* 1 1
Robert Thornton, Esq. M.P. F.A.S. P.R.I. 5 5
Samuel Thornton, Esq. F.A.S. P.R.I. 1 1
Samuel Thornton, Esq. 1 1
Robert John Thornton, M.D. 1 1
Anthony Thorp, Esq. *York* 1 1
William Hall Timbrel, Esq. 1 1
Mr. Cornelius Tongue 1 1
John Townshend, Esq. *Chelsea* 5 5
John Townsend, Esq. *Wandsworth* 5 5
Samuel Turner, Esq. 2 2
Thomas Turner, M.D. 2 2
Sharon Turner, Esq. F.A.S. 1 1
Dawson Turner, Esq. F.A.S. 2 2
J. Turton, Esq. 1 1
Thomas Tyndall, Esq. *Bristol*

U
Peter Vere, Esq. F.A.S. 1 1
Mr. Underwood 1 1
James Upton, Esq. 1 1
J. Upton, Esq. 1 1
T. Upward, Esq. 2 2
Miss S. Vranken, *Bristol*

Gordon Urquhart, Esq. 1 1

W
Sir William E. Welby, Bart. 2 2
Honourable J. S. Wortley, P.R.I. 1 1
M. F. Wagstaff, Esq. 1 1
Daniel Walker, Esq. 1 1
Samuel Walker, Esq. *Masbrough*
Mr. J. Walker 1 1
Mr. Joseph Walker 1 1
Martin Wall, M.D. *Clinical Professor, Oxford*
T. Walshman, M.D. _Physician to the Western and Surry Dispensaries, and President of the Physical Society 1 1
John Walter, Esq. *Pimlico* 1 1
D. Walters, Esq. 1 1
Olaus Warberg, Esq. *Copenhagen* 1 1
Mrs. Ward, *Hampstead* 1 1
Gilbert Wardlaw, Esq. *Glasgow* 1 1
James Ware, Esq. F.R.S. F.A.S. 5 5
Rev. John Ware, *York*
John Warren, Esq. P.R.I. 1 1
Mr. Watkins 2 2
Mr. Watson 2 2
W. H. Watts, Esq. 5 5
David Pike Watts, Esq. P.R.I. 2 2
Mrs. Wawn 1 1
Rev. George Wearing, *Whalley* 1 1
Mr. James Weatherby 1 1
Mr. Matthew Wedeson 1 1
J. Wedgwood, Esq. P.R.I. 1 1
J. C. Weguelin, Esq. 2 2
Mr. J. Welbank 1 1
William Earle Welby, Esq. *Carlton House, Nottinghamshire* 3 3
Richard Earle Welby, Esq. 2 2
George Welch, Esq. 1 1
Mrs. Welch 2 2

Thomas West, Esq. *Bath* 1 1
Mrs. West 5 5
Westminster Library 1 1
Lawson Whalley, Esq. *Lancaster* 1 1
Lawson Whalley, Esq. *Lancaster* 1 1
The Rev. John Wheeler 1 1
Samuel Whitbread, Esq. M.P. 3 3
Edward White, Esq. 1 1
Rev. Samuel White 1 1
Mrs. White, *Bury St. Edmunds* 2 2
Miss White 1 1
Rev. Walter Whiter 1 1
Robert Whitfield, Esq. *Surgeon* 1 1
William Whittington, Esq. 5 5
Miss Whittington 3 3
Miss E. Whittington 2 2
Edward Athenry Whyte, Esq. *Dublin* 1 1
Rev. Thomas Wigan 1 1
Mr. John Wiglesworth 1 1
Lieut. Col. Wilder 1 1
Mr. Wilkes 1 1
Thomas Willan, Esq. 5 5
Mrs. Willlan 5 5
John Willan, Esq. 5 5
John Williams, Esq. *Caermarthen, Surgeon* 1 1
— — —Williams, Esq. 1 1
Mr. William Henry Williams, *Caius College, Cambridge* 1 1
Dr. A. F. M. Willich 1 1
Dr. Robert Willis 5 5
H. W. C. Wilson, Esq. 2 2
James Wilson, Esq. 2 2
Dr. Wilson 2 2
Rev. Thomas Wilson, *Clitheroe* 1 1
Miss Winckworth 1 1
Joseph Windham, Esq. F.R.S. F.L.S. 2 2
Alexander Wood, Esq. *Edinburgh, Surgeon*

R. Wood, Esq. 1 1
Robert Wood, Esq. *Tamworth* 1 1
William Wood, Esq. *Hackney* 1 1
Mr. James Wood 1 1
Samuel Woods, Esq. 3 3
Mr. Henry Worboys 1 1
Mr. John Worboys 1 1
Peter Wright, Esq. P.R.I. 2 2
Mr. James Wright, *Liverpool* 2 2
Dr. Wright 1 1
Mr. Wright, *Optician* 1 1
William Wynch, Esq. P.R.I. 2 2
W. Wynch, Esq. *2nd Subscription* 1 1

Y
Joseph Yallowley, Esq. 1 1
Rev. John Yates 3 3
Matthew Yatman, Esq. F.A.S. 1 1
Dr. Yelloly 2 2
J. W. Yonge, Esq. 2 2
Thomas Young, M.D. For. Sec. R.S.
    F.L.S. 5 5
G. Young, Esq. 1 1
William Younge, M.D. *Sheffield*

From the Press of the Royal Institution of Great Britain, Albemarle Street, London: William Savage, Printer.

ADDENDA.
L. s.
A
The Right Hon. Lord Altamont 1 1
Nicholas Ashton, Esq. 5 5
J. J. Angerstein, Esq. 10 10
Mr. James Asperne 1 1
J. B. Austin, Esq. 1 1
Mr. Abernethy 5 5

B
Scrope Barnard, Esq. 1 1
C. F. Bamwell, Esq. 1 1
Rev. Dr. Balward 1 1
Rev. T. Browne 5 5
Rev. B. Bridge 1 1
Rev. T. Brooke 1 1
W. S. Bruere, Esq. 1 1
J. D. Borton, Esq. 1 1
William Breese, Esq. 2 2
Mr. Bradner 1 1
James Bureau, Esq. 3 3
Mrs. Mary Bowyer 1 1
Rev. G. Butler 1 1
Rev. J. Brown 1 1
Mr. Bryant 1 1
John Butts, Esq. 1 1
Mr. Bruce Boswell 1 1

C
James Collins, Esq. 1 1
William Cass, Esq. 2 2
M. Da Costa 1 1
F. Carter, Esq. 1 1
Rev. T. Cautley 1 1
G. Caldwell, Esq. 1 1
Rev. T. Clarkson 1 1
E. D. Clarke, Esq. 1 1
T. M. Cripps, Esq. 1 1
Rev. B. Chapman 1 1

D
The Lord Bishop of Durham 10 10
P. Denys, Esq. 4 4
Rev. F. Drake 1 1
J. Dearman, Esq. *Champion Hill* 1 1
M. Dobson, Esq. 1 1
Dr. Denman 2 2
Mr. Durham 2 0

Rev. Dr. Dealtry 1 1
Martyn Davy, M.D. 1 1
Philip Derbishire, Esq. 1 1

E
Sir James Esdaile 1 1
Dr. G. B. Eaton 1 1

F
John Fuller, Esq. 1 1
Edward Foster, Esq. 1 1
Edward Foster, jun. Esq. 1 1
John Fellowes, Esq. 1 1
A friend, W. H. S. by Mr. Cuppage 1 1
A friend 5 5
A friend, S. B. 2 2
A friend, R. R. 1 1
A friend, C. A. Esq. 1 1

G
Stephen Gaselee, Esq. 2 2
W. B. Garlike, M.D. 2 2
D. Garnett, Esq. 1 1
Mrs. Garnett 1 1
J. R. Gladdow, Esq. 1 1

H
The Hon. Major Henniker 1 1
J. Hall, Esq. 1 1
– – – Hill, Esq. 2 2
Richard Hussey, Esq. 2 2
Mark Huish, jun. Esq. 1 1
Nathaniel Hubbins, Esq. 1 1
Dr. Heberden 2 2
T. Holmes, Esq. 1 1
John Holkyard, Esq. *Surgeon* 1 1
Professor Harwood 2 2
E. Hibgame, Esq. 1 1
T. P. Hackhouse, Esq. 1 1

B. Harvey, Esq. 1 1
Rev. T. Holden 1 1
Rev. Mr. Holme 1 1
William Hutton, Esq. 1 1
Dr. Halliday 2 2
Thomas Halliday, Esq. 1 1
William Holden, Esq. 1 1
— — —Van Hoorst, Esq. 1 1

J
Rev. T. Jones 1 1
Mr. S. Jones 1 1

K
L. Knight, Esq. 1 1
Rev. J. Kenrick 1 1

L
Sir John Leggart, Bart. 1 1
William Lewis, Esq. 1 1
A. S. Lillingstone, Esq. 2 2
Dr. Latham 5 5
M

Basil Montague, Esq. 1 1
James Mackintosh, Esq. 1 1
H. Mayers, Esq. 1 1
Lieutenant Col. McDonald 1 1
M. Marshall, Esq. 1 1
Rev. Mr. Maule 1 1
Rev. Mr. Manning 1 1
Rev. W. Miller 1 1
Dr. John William Minier 2 2
Mr. Miller 1 1
Mr. Marshall 1 1

N
Richard Nixon, Esq. 1 1
George Norman, Esq. 3 3

— — —Nottage, Esq. 2 2
Mr. Richard Nicholls 1 1

O
Benjamin Oakley, Esq. 1 1
Rev. W. Otter 1 1

P
Hon. Mr. Pakenham 1 1
N. Palmer, Esq. 1 1
Dr. David Pitcairn 5 5
Dr. Pennington 1 1
The Rev. G. Pollen 10 10
St. Peter's College 6 6

R
W. Robinson, Esq. 1 1
William Rathbone, Esq. 5 5
Rev. H. R. Rouney 1 1
Rev. Mr. Renouard 1 1
Mr. Rawlins 1 1

S
A. Smith, Esq. 2 2
James Skirrow, Esq. 1 1
Dr. Sloper 5 5
John Simpson, M.D. 1 1
Rev. Mr. Simeon 1 1
Rev. J. Satterthwaite 1 1

T
The Right Hon. John Trevor 2 2
Rev. Dr. Towitt 1 1
Mr. Thomas Teasdale 1 1

V
Rev. T. Vickers 1 1

W
M. S. Wakefield, Esq. 1 1
Thomas Wilson, Esq. 2 2
R. S. Wells, Esq. 1 1
William White, Esq. 1 1
Joseph Walker, Esq. 1 1
John Welbank, Esq. 1 1
Rev. J. Wood 1 1
Rev. Mr. Wheatear 1 1
Rev. J. Walker 1 1
Mrs. Ward 1 1

Y
J. A. Young, Esq. 2 2
Dr. Young 1 1